Use R!

Series Editors:
Robert Gentleman   Kurt Hornik   Giovanni Parmigiani

Brian Everitt • Torsten Hothorn

An Introduction to Applied Multivariate Analysis with R

Springer

Brian Everitt
Professor Emeritus
King's College
London, SE5 8AF
UK
brian.everitt@btopenworld.com

Torsten Hothorn
Institut für Statistik
Ludwig-Maximilians-Universität München
Ludwigstr. 33
80539 München
Germany
Torsten.Hothorn@stat.uni-muenchen.de

Series Editors:
Robert Gentleman
Program in Computational Biology
Division of Public Health Sciences
Fred Hutchinson Cancer Research Center
1100 Fairview Avenue, N. M2-B876
Seattle, Washington 98109
USA

Kurt Hornik
Department of Statistik and Mathematik
Wirtschaftsuniversität Wien
Augasse 2-6
A-1090 Wien
Austria

Giovanni Parmigiani
The Sidney Kimmel Comprehensive
Cancer Center at Johns Hopkins University
550 North Broadway
Baltimore, MD 21205-2011
USA

ISBN 978-1-4419-9649-7   e-ISBN 978-1-4419-9650-3
DOI 10.1007/978-1-4419-9650-3
Springer New York Dordrecht Heidelberg London

Library of Congress Control Number: 2011926793

Printed on acid-free paper

Springer is part of Springer Science+Business Media (www.springer.com)

To our wives, Mary-Elizabeth and Carolin.

Preface

The majority of data sets collected by researchers in all disciplines are multivariate, meaning that several measurements, observations, or recordings are taken on each of the units in the data set. These units might be human subjects, archaeological artifacts, countries, or a vast variety of other things. In a few cases, it may be sensible to isolate each variable and study it separately, but in most instances all the variables need to be examined simultaneously in order to fully grasp the structure and key features of the data. For this purpose, one or another method of multivariate analysis might be helpful, and it is with such methods that this book is largely concerned. Multivariate analysis includes methods both for describing and exploring such data and for making formal inferences about them. The aim of all the techniques is, in a general sense, to display or extract the signal in the data in the presence of noise and to find out what the data show us in the midst of their apparent chaos.

The computations involved in applying most multivariate techniques are considerable, and their routine use requires a suitable software package. In addition, most analyses of multivariate data should involve the construction of appropriate graphs and diagrams, and this will also need to be carried out using the same package. R is a statistical computing environment that is powerful, flexible, and, in addition, has excellent graphical facilities. It is for these reasons that it is the use of R for multivariate analysis that is illustrated in this book.

In this book, we concentrate on what might be termed the "core" or "classical" multivariate methodology, although mention will be made of recent developments where these are considered relevant and useful. But there is an area of multivariate statistics that we have omitted from this book, and that is multivariate analysis of variance (MANOVA) and related techniques such as Fisher's linear discriminant function (LDF). There are a variety of reasons for this omission. First, we are not convinced that MANOVA is now of much more than historical interest; researchers may occasionally pay lip service to using the technique, but in most cases it really is no more than this. They quickly move on to looking at the results for individual variables. And MANOVA for repeated measures has been largely superseded by the models that we shall describe in Chapter 8. Second, a classification technique such as LDF needs to be considered in the context of modern classification algorithms, and these cannot be covered in an introductory book such as this.

Some brief details of the theory behind each technique described are given, but the main concern of each chapter is the correct application of the methods so as to extract as much information as possible from the data at hand, particularly as some type of graphical representation, via the R software.

The book is aimed at students in applied statistics courses, both undergraduate and post-graduate, who have attended a good introductory course in statistics that covered hypothesis testing, confidence intervals, simple regression and correlation, analysis of variance, and basic maximum likelihood estimation. We also assume that readers will know some simple matrix algebra, including the manipulation of matrices and vectors and the concepts of the inverse and rank of a matrix. In addition, we assume that readers will have some familiarity with R at the level of, say, Dalgaard (2002). In addition to such a student readership, we hope that many applied statisticians dealing with multivariate data will find something of interest in the eight chapters of our book.

Throughout the book, we give many examples of R code used to apply the multivariate techniques to multivariate data. Samples of code that could be entered interactively at the R command line are formatted as follows:

```
R> library("MVA")
```

Here, R> denotes the prompt sign from the R command line, and the user enters everything else. The symbol + indicates additional lines, which are appropriately indented. Finally, output produced by function calls is shown below the associated code:

```
R> rnorm(10)

 [1]  1.8808  0.2572 -0.3412  0.4081  0.4344  0.7003  1.8944
 [8] -0.2993 -0.7355  0.8960
```

In this book, we use several R packages to access different example data sets (many of them contained in the package **HSAUR2**), standard functions for the general parametric analyses, and the **MVA** package to perform analyses. All of the packages used in this book are available at the Comprehensive R Archive Network (CRAN), which can be accessed from http://CRAN.R-project.org.

The source code for the analyses presented in this book is available from the **MVA** package. A demo containing the R code to reproduce the individual results is available for each chapter by invoking

```
R> library("MVA")
R> demo("Ch-MVA") ### Introduction to Multivariate Analysis
R> demo("Ch-Viz") ### Visualization
```

```
R> demo("Ch-PCA") ### Principal Components Analysis
R> demo("Ch-EFA") ### Exploratory Factor Analysis
R> demo("Ch-MDS") ### Multidimensional Scaling
R> demo("Ch-CA")  ### Cluster Analysis
R> demo("Ch-SEM") ### Structural Equation Models
R> demo("Ch-LME") ### Linear Mixed-Effects Models
```

Thanks are due to Lisa Möst, BSc., for help with data processing and LATEX typesetting, the copy editor for many helpful corrections, and to John Kimmel, for all his support and patience during the writing of the book.

January 2011
Brian S. Everitt, London
Torsten Hothorn, München

Contents

1

Multivariate Data and Multivariate Analysis

1.1 Introduction

Multivariate data arise when researchers record the values of several random variables on a number of subjects or objects or perhaps one of a variety of other things (we will use the general term "units") in which they are interested, leading to a *vector-valued* or *multidimensional* observation for each. Such data are collected in a wide range of disciplines, and indeed it is probably reasonable to claim that the majority of data sets met in practise are multivariate. In some studies, the variables are chosen by design because they are known to be essential descriptors of the system under investigation. In other studies, particularly those that have been difficult or expensive to organise, many variables may be measured simply to collect as much information as possible as a matter of expediency or economy.

Multivariate data are ubiquitous as is illustrated by the following four examples:

- Psychologists and other behavioural scientists often record the values of several different cognitive variables on a number of subjects.
- Educational researchers may be interested in the examination marks obtained by students for a variety of different subjects.
- Archaeologists may make a set of measurements on artefacts of interest.
- Environmentalists might assess pollution levels of a set of cities along with noting other characteristics of the cities related to climate and human ecology.

Most multivariate data sets can be represented in the same way, namely in a rectangular format known from spreadsheets, in which the elements of each row correspond to the variable values of a particular unit in the data set and the elements of the columns correspond to the values taken by a particular variable. We can write data in such a rectangular format as

Unit	Variable 1	...	Variable q
1	x_{11}	$\cdots$	x_{1q}
$\vdots$	$\vdots$	$\vdots$	$\vdots$
n	x_{n1}	$\cdots$	x_{nq}

where n is the number of units, q is the number of variables recorded on each unit, and x_{ij} denotes the value of the jth variable for the ith unit. The observation part of the table above is generally represented by an $n \times q$ *data matrix*, $\mathbf{X}$. In contrast to the *observed* data, the theoretical entities describing the univariate distributions of each of the q variables and their joint distribution are denoted by so-called *random variables* $X_1, \ldots, X_q$.

Although in some cases where multivariate data have been collected it may make sense to isolate each variable and study it separately, in the main it does not. Because the whole set of variables is measured on *each* unit, the variables will be related to a greater or lesser degree. Consequently, if each variable is analysed in isolation, the full structure of the data may not be revealed. *Multivariate statistical analysis* is the *simultaneous* statistical analysis of a collection of variables, which improves upon separate univariate analyses of each variable by using information about the *relationships* between the variables. Analysis of each variable separately is very likely to miss uncovering the key features of, and any interesting "patterns" in, the multivariate data.

The units in a set of multivariate data are sometimes sampled from a population of interest to the investigator, a population about which he or she wishes to make some inference or other. More often perhaps, the units cannot really be said to have been sampled from some population in any meaningful sense, and the questions asked about the data are then largely exploratory in nature. with the ubiquitous p-value of univariate statistics being notable by its absence. Consequently, there are methods of multivariate analysis that are essentially exploratory and others that can be used for statistical inference.

For the exploration of multivariate data, formal models designed to yield specific answers to rigidly defined questions are not required. Instead, methods are used that allow the detection of possibly unanticipated patterns in the data, opening up a wide range of competing explanations. Such methods are generally characterised both by an emphasis on the importance of graphical displays and visualisation of the data and the lack of any associated probabilistic model that would allow for formal inferences. Multivariate techniques that are largely exploratory are described in Chapters 2 to 6.

A more formal analysis becomes possible in situations when it is realistic to assume that the individuals in a multivariate data set have been sampled from some population and the investigator wishes to test a well-defined hypothesis about the parameters of that population's probability density function. Now the main focus will not be the sample data per se, but rather on using information gathered from the sample data to draw inferences about the population. And the probability density function almost universally assumed as the basis of inferences for multivariate data is the *multivariate normal*. (For a brief description of the multivariate normal density function and ways of assessing whether a set of multivariate data conform to the density, see Section 1.6). Multivariate techniques for which formal inference is of importance are described in Chapters 7 and 8. But in many cases when dealing with multivariate data, this implied distinction between the exploratory and the inferential may be a red herring because the general aim of *most* multivariate analyses, whether implicitly exploratory or inferential is to uncover, display, or extract any "signal" in the data in the presence of noise and to discover what the data have to tell us.

1.2 A brief history of the development of multivariate analysis

The genesis of multivariate analysis is probably the work carried out by Francis Galton and Karl Pearson in the late 19th century on quantifying the relationship between offspring and parental characteristics and the development of the correlation coefficient. And then, in the early years of the 20th century, Charles Spearman laid down the foundations of factor analysis (see Chapter 5) whilst investigating correlated intelligence quotient (IQ) tests. Over the next two decades, Spearman's work was extended by Hotelling and by Thurstone.

Multivariate methods were also motivated by problems in scientific areas other than psychology, and in the 1930s Fisher developed linear discriminant function analysis to solve a taxonomic problem using multiple botanical measurements. And Fisher's introduction of analysis of variance in the 1920s was soon followed by its multivariate generalisation, multivariate analysis of variance, based on work by Bartlett and Roy. (These techniques are not covered in this text for the reasons set out in the Preface.)

In these early days, computational aids to take the burden of the vast amounts of arithmetic involved in the application of the multivariate methods being proposed were very limited and, consequently, developments were primarily mathematical and multivariate research was, at the time, largely a branch of linear algebra. However, the arrival and rapid expansion of the use of electronic computers in the second half of the 20th century led to increased practical application of existing methods of multivariate analysis and renewed interest in the creation of new techniques.

In the early years of the 21st century, the wide availability of relatively cheap and extremely powerful personal computers and laptops allied with flexible statistical software has meant that all the methods of multivariate analysis can be applied routinely even to very large data sets such as those generated in, for example, genetics, imaging, and astronomy. And the application of multivariate techniques to such large data sets has now been given its own name, *data mining*, which has been defined as "the nontrivial extraction of implicit, previously unknown and potentially useful information from data." Useful books on data mining are those of Fayyad, Piatetsky-Shapiro, Smyth, and Uthurusamy (1996) and Hand, Mannila, and Smyth (2001).

1.3 Types of variables and the possible problem of missing values

A hypothetical example of multivariate data is given in Table 1.1. The special symbol NA denotes missing values (being Not Available); the value of this variable for a subject is missing.

Table 1.1: **hypo** data. Hypothetical Set of Multivariate Data.

individual	sex	age	IQ	depression	health	weight
1	Male	21	120	Yes	Very good	150
2	Male	43	NA	No	Very good	160
3	Male	22	135	No	Average	135
4	Male	86	150	No	Very poor	140
5	Male	60	92	Yes	Good	110
6	Female	16	130	Yes	Good	110
7	Female	NA	150	Yes	Very good	120
8	Female	43	NA	Yes	Average	120
9	Female	22	84	No	Average	105
10	Female	80	70	No	Good	100

Here, the number of units (people in this case) is $n = 10$, with the number of variables being $q = 7$ and, for example, $x_{34} = 135$. In R, a "**data.frame**" is the appropriate data structure to represent such rectangular data. Subsets of units (rows) or variables (columns) can be extracted via the [subset operator; i.e.,

```
R> hypo[1:2, c("health", "weight")]
```

```
     health weight
1 Very good    150
2 Very good    160
```

extracts the values x_{15}, x_{16} and x_{25}, x_{26} from the hypothetical data presented in Table 1.1. These data illustrate that the variables that make up a set of multivariate data will not necessarily all be of the same type. Four levels of measurements are often distinguished:

Nominal: Unordered categorical variables. Examples include treatment allocation, the sex of the respondent, hair colour, presence or absence of depression, and so on.

Ordinal: Where there is an ordering but no implication of equal distance between the different points of the scale. Examples include social class, self-perception of health (each coded from I to V, say), and educational level (no schooling, primary, secondary, or tertiary education).

Interval: Where there are equal differences between successive points on the scale but the position of zero is arbitrary. The classic example is the measurement of temperature using the Celsius or Fahrenheit scales.

Ratio: The highest level of measurement, where one can investigate the *relative magnitudes* of scores as well as the differences between them. The position of zero is fixed. The classic example is the absolute measure of temperature (in Kelvin, for example), but other common ones includes age (or any other time from a fixed event), weight, and length.

In many statistical textbooks, discussion of different types of measurements is often followed by recommendations as to which statistical techniques are suitable for each type; for example, analyses on nominal data should be limited to summary statistics such as the number of cases, the mode, etc. And, for ordinal data, means and standard deviations are not suitable. But Velleman and Wilkinson (1993) make the important point that restricting the choice of statistical methods in this way may be a dangerous practise for data analysis–in essence the measurement taxonomy described is often too strict to apply to real-world data. This is not the place for a detailed discussion of measurement, but we take a fairly pragmatic approach to such problems. For example, we will not agonise over treating variables such as measures of depression, anxiety, or intelligence as if they are interval-scaled, although strictly they fit into the ordinal category described above.

1.3.1 Missing values

Table 1.1 also illustrates one of the problems often faced by statisticians undertaking statistical analysis in general and multivariate analysis in particular, namely the presence of *missing values* in the data; i.e., observations and measurements that should have been recorded but for one reason or another, were not. Missing values in multivariate data may arise for a number of reasons; for example, non-response in sample surveys, dropouts in *longitudinal data* (see Chapter 8), or refusal to answer particular questions in a questionnaire. The most important approach for dealing with missing data is to try to avoid them during the data-collection stage of a study. But despite all the efforts a researcher may make, he or she may still be faced with a data set that contains a number of missing values. So what can be done? One answer to this question is to take the *complete-case analysis* route because this is what most statistical software packages do automatically. Using complete-case analysis on multivariate data means omitting *any* case with a missing value on any of the variables. It is easy to see that if the number of variables is large, then even a sparse pattern of missing values can result in a substantial number of incomplete cases. One possibility to ease this problem is to simply drop any variables that have many missing values. But complete-case analysis is not recommended for two reasons:

- Omitting a possibly substantial number of individuals will cause a large amount of information to be discarded and lower the effective sample size of the data, making any analyses less effective than they would have been if all the original sample had been available.
- More worrisome is that dropping the cases with missing values on one or more variables can lead to serious biases in both estimation and inference unless the discarded cases are essentially a random subsample of the observed data (the term *missing completely at random* is often used; see Chapter 8 and Little and Rubin (1987) for more details).

So, at the very least, complete-case analysis leads to a loss, and perhaps a substantial loss, in power by discarding data, but worse, analyses based just on complete cases might lead to misleading conclusions and inferences.

A relatively simple alternative to complete-case analysis that is often used is *available-case analysis*. This is a straightforward attempt to exploit the incomplete information by using all the cases available to estimate quantities of interest. For example, if the researcher is interested in estimating the *correlation matrix* (see Subsection 1.5.2) of a set of multivariate data, then available-case analysis uses all the cases with variables X_i and X_j present to estimate the correlation between the two variables. This approach appears to make better use of the data than complete-case analysis, but unfortunately available-case analysis has its own problems. The sample of individuals used changes from correlation to correlation, creating potential difficulties when the missing data are not missing completely at random. There is no guarantee that the estimated correlation matrix is even positive-definite which can create problems for some of the methods, such as *factor analysis* (see Chapter 5) and *structural equation modelling* (see Chapter 7), that the researcher may wish to apply to the matrix.

Both complete-case and available-case analyses are unattractive unless the number of missing values in the data set is "small". An alternative answer to the missing-data problem is to consider some form of *imputation*, the practise of "filling in" missing data with plausible values. Methods that impute the missing values have the advantage that, unlike in complete-case analysis, observed values in the incomplete cases are retained. On the surface, it looks like imputation will solve the missing-data problem and enable the investigator to progress normally. But, from a statistical viewpoint, careful consideration needs to be given to the *method* used for imputation or otherwise it may cause more problems than it solves; for example, imputing an observed variable mean for a variable's missing values preserves the observed sample means but distorts the *covariance matrix* (see Subsection 1.5.1), biasing estimated variances and covariances towards zero. On the other hand, imputing predicted values from regression models tends to inflate observed correlations, biasing them away from zero (see Little 2005). And treating imputed data as if they were "real" in estimation and inference can lead to misleading standard errors and p-values since they fail to reflect the uncertainty due to the missing data.

The most appropriate way to deal with missing values is by a procedure suggested by Rubin (1987) known as *multiple imputation*. This is a Monte Carlo technique in which the missing values are replaced by $m > 1$ simulated versions, where m is typically small (say 3–10). Each of the simulated complete data sets is analysed using the method appropriate for the investigation at hand, and the results are later combined to produce, say, estimates and confidence intervals that incorporate missing-data uncertainty. Details are given in Rubin (1987) and more concisely in Schafer (1999). The great virtues of multiple imputation are its simplicity and its generality. The user may analyse the data using virtually any technique that would be appropriate if the data were complete. However, one should always bear in mind that the imputed values are not real measurements. We do not get something for nothing! And if there is a substantial proportion of individuals with large amounts of missing data, one should clearly question whether *any* form of statistical analysis is worth the bother.

1.4 Some multivariate data sets

This is a convenient point to look at some multivariate data sets and briefly ponder the type of question that might be of interest in each case. The first data set consists of chest, waist, and hip measurements on a sample of men and women and the measurements for 20 individuals are shown in Table 1.2. Two questions might be addressed by such data;

- Could body size and body shape be summarised in some way by combining the three measurements into a single number?
- Are there subtypes of body shapes amongst the men and amongst the women within which individuals are of similar shapes and between which body shapes differ?

The first question might be answered by *principal components analysis* (see Chapter 3), and the second question could be investigated using *cluster analysis* (see Chapter 6).

(In practise, it seems intuitively likely that we would have needed to record the three measurements on many more than 20 individuals to have any chance of being able to get convincing answers from these techniques to the questions of interest. The question of how many units are needed to achieve a sensible analysis when using the various techniques of multivariate analysis will be taken up in the respective chapters describing each technique.)

Table 1.2: **measure** data. Chest, waist, and hip measurements on 20 individuals (in inches).

chest	waist	hips	gender	chest	waist	hips	gender
34	30	32	male	36	24	35	female
37	32	37	male	36	25	37	female
38	30	36	male	34	24	37	female
36	33	39	male	33	22	34	female
38	29	33	male	36	26	38	female
43	32	38	male	37	26	37	female
40	33	42	male	34	25	38	female
38	30	40	male	36	26	37	female
40	30	37	male	38	28	40	female
41	32	39	male	35	23	35	female

Our second set of multivariate data consists of the results of chemical analysis on Romano-British pottery made in three different regions (region 1 contains kiln 1, region 2 contains kilns 2 and 3, and region 3 contains kilns 4 and 5). The complete data set, which we shall meet in Chapter 6, consists of the chemical analysis results on 45 pots, shown in Table 1.3. One question that might be posed about these data is whether the chemical profiles of each pot suggest different types of pots and if any such types are related to kiln or region. This question is addressed in Chapter 6.

Table 1.3: **pottery** data. Romano-British pottery data.

Al2O3	Fe2O3	MgO	CaO	Na2O	K2O	TiO2	MnO	BaO	kiln
18.8	9.52	2.00	0.79	0.40	3.20	1.01	0.077	0.015	1
16.9	7.33	1.65	0.84	0.40	3.05	0.99	0.067	0.018	1
18.2	7.64	1.82	0.77	0.40	3.07	0.98	0.087	0.014	1
16.9	7.29	1.56	0.76	0.40	3.05	1.00	0.063	0.019	1
17.8	7.24	1.83	0.92	0.43	3.12	0.93	0.061	0.019	1
18.8	7.45	2.06	0.87	0.25	3.26	0.98	0.072	0.017	1
16.5	7.05	1.81	1.73	0.33	3.20	0.95	0.066	0.019	1
18.0	7.42	2.06	1.00	0.28	3.37	0.96	0.072	0.017	1
15.8	7.15	1.62	0.71	0.38	3.25	0.93	0.062	0.017	1
14.6	6.87	1.67	0.76	0.33	3.06	0.91	0.055	0.012	1
13.7	5.83	1.50	0.66	0.13	2.25	0.75	0.034	0.012	1
14.6	6.76	1.63	1.48	0.20	3.02	0.87	0.055	0.016	1
14.8	7.07	1.62	1.44	0.24	3.03	0.86	0.080	0.016	1
17.1	7.79	1.99	0.83	0.46	3.13	0.93	0.090	0.020	1
16.8	7.86	1.86	0.84	0.46	2.93	0.94	0.094	0.020	1
15.8	7.65	1.94	0.81	0.83	3.33	0.96	0.112	0.019	1

Table 1.3: **pottery** data (continued).

Al2O3	Fe2O3	MgO	CaO	Na2O	K2O	TiO2	MnO	BaO	kiln
18.6	7.85	2.33	0.87	0.38	3.17	0.98	0.081	0.018	1
16.9	7.87	1.83	1.31	0.53	3.09	0.95	0.092	0.023	1
18.9	7.58	2.05	0.83	0.13	3.29	0.98	0.072	0.015	1
18.0	7.50	1.94	0.69	0.12	3.14	0.93	0.035	0.017	1
17.8	7.28	1.92	0.81	0.18	3.15	0.90	0.067	0.017	1
14.4	7.00	4.30	0.15	0.51	4.25	0.79	0.160	0.019	2
13.8	7.08	3.43	0.12	0.17	4.14	0.77	0.144	0.020	2
14.6	7.09	3.88	0.13	0.20	4.36	0.81	0.124	0.019	2
11.5	6.37	5.64	0.16	0.14	3.89	0.69	0.087	0.009	2
13.8	7.06	5.34	0.20	0.20	4.31	0.71	0.101	0.021	2
10.9	6.26	3.47	0.17	0.22	3.40	0.66	0.109	0.010	2
10.1	4.26	4.26	0.20	0.18	3.32	0.59	0.149	0.017	2
11.6	5.78	5.91	0.18	0.16	3.70	0.65	0.082	0.015	2
11.1	5.49	4.52	0.29	0.30	4.03	0.63	0.080	0.016	2
13.4	6.92	7.23	0.28	0.20	4.54	0.69	0.163	0.017	2
12.4	6.13	5.69	0.22	0.54	4.65	0.70	0.159	0.015	2
13.1	6.64	5.51	0.31	0.24	4.89	0.72	0.094	0.017	2
11.6	5.39	3.77	0.29	0.06	4.51	0.56	0.110	0.015	3
11.8	5.44	3.94	0.30	0.04	4.64	0.59	0.085	0.013	3
18.3	1.28	0.67	0.03	0.03	1.96	0.65	0.001	0.014	4
15.8	2.39	0.63	0.01	0.04	1.94	1.29	0.001	0.014	4
18.0	1.50	0.67	0.01	0.06	2.11	0.92	0.001	0.016	4
18.0	1.88	0.68	0.01	0.04	2.00	1.11	0.006	0.022	4
20.8	1.51	0.72	0.07	0.10	2.37	1.26	0.002	0.016	4
17.7	1.12	0.56	0.06	0.06	2.06	0.79	0.001	0.013	5
18.3	1.14	0.67	0.06	0.05	2.11	0.89	0.006	0.019	5
16.7	0.92	0.53	0.01	0.05	1.76	0.91	0.004	0.013	5
14.8	2.74	0.67	0.03	0.05	2.15	1.34	0.003	0.015	5
19.1	1.64	0.60	0.10	0.03	1.75	1.04	0.007	0.018	5

Source: Tubb, A., et al., *Archaeometry*, 22, 153–171, 1980. With permission.

Our third set of multivariate data involves the examination scores of a large number of college students in six subjects; the scores for five subjects are shown in Table 1.4. Here the main question of interest might be whether the exam scores reflect some underlying trait in a student that cannot be measured directly, perhaps "general intelligence"? The question could be investigated by using *exploratory factor analysis* (see Chapter 5).

Table 1.4: **exam** data. Exam scores for five psychology students.

subject	maths	english	history	geography	chemistry	physics
1	60	70	75	58	53	42
2	80	65	66	75	70	76
3	53	60	50	48	45	43
4	85	79	71	77	68	79
5	45	80	80	84	44	46

The final set of data we shall consider in this section was collected in a study of air pollution in cities in the USA. The following variables were obtained for 41 US cities:

SO2: SO_2 content of air in micrograms per cubic metre;
temp: average annual temperature in degrees Fahrenheit;
manu: number of manufacturing enterprises employing 20 or more workers;
popul: population size (1970 census) in thousands;
wind: average annual wind speed in miles per hour;
precip: average annual precipitation in inches;
predays: average number of days with precipitation per year.

The data are shown in Table 1.5.

Table 1.5: **USairpollution** data. Air pollution in 41 US cities.

	SO2	temp	manu	popul	wind	precip	predays
Albany	46	47.6	44	116	8.8	33.36	135
Albuquerque	11	56.8	46	244	8.9	7.77	58
Atlanta	24	61.5	368	497	9.1	48.34	115
Baltimore	47	55.0	625	905	9.6	41.31	111
Buffalo	11	47.1	391	463	12.4	36.11	166
Charleston	31	55.2	35	71	6.5	40.75	148
Chicago	110	50.6	3344	3369	10.4	34.44	122
Cincinnati	23	54.0	462	453	7.1	39.04	132
Cleveland	65	49.7	1007	751	10.9	34.99	155
Columbus	26	51.5	266	540	8.6	37.01	134
Dallas	9	66.2	641	844	10.9	35.94	78
Denver	17	51.9	454	515	9.0	12.95	86
Des Moines	17	49.0	104	201	11.2	30.85	103
Detroit	35	49.9	1064	1513	10.1	30.96	129
Hartford	56	49.1	412	158	9.0	43.37	127
Houston	10	68.9	721	1233	10.8	48.19	103
Indianapolis	28	52.3	361	746	9.7	38.74	121
Jacksonville	14	68.4	136	529	8.8	54.47	116

Table 1.5: USairpollution data (continued).

	SO2	temp	manu	popul	wind	precip	predays
Kansas City	14	54.5	381	507	10.0	37.00	99
Little Rock	13	61.0	91	132	8.2	48.52	100
Louisville	30	55.6	291	593	8.3	43.11	123
Memphis	10	61.6	337	624	9.2	49.10	105
Miami	10	75.5	207	335	9.0	59.80	128
Milwaukee	16	45.7	569	717	11.8	29.07	123
Minneapolis	29	43.5	699	744	10.6	25.94	137
Nashville	18	59.4	275	448	7.9	46.00	119
New Orleans	9	68.3	204	361	8.4	56.77	113
Norfolk	31	59.3	96	308	10.6	44.68	116
Omaha	14	51.5	181	347	10.9	30.18	98
Philadelphia	69	54.6	1692	1950	9.6	39.93	115
Phoenix	10	70.3	213	582	6.0	7.05	36
Pittsburgh	61	50.4	347	520	9.4	36.22	147
Providence	94	50.0	343	179	10.6	42.75	125
Richmond	26	57.8	197	299	7.6	42.59	115
Salt Lake City	28	51.0	137	176	8.7	15.17	89
San Francisco	12	56.7	453	716	8.7	20.66	67
Seattle	29	51.1	379	531	9.4	38.79	164
St. Louis	56	55.9	775	622	9.5	35.89	105
Washington	29	57.3	434	757	9.3	38.89	111
Wichita	8	56.6	125	277	12.7	30.58	82
Wilmington	36	54.0	80	80	9.0	40.25	114

Source: Sokal, R. R., Rohlf, F. J., *Biometry*, W. H. Freeman, San Francisco, 1981. With permission.

What might be the question of most interest about these data? Very probably it is "how is pollution level as measured by sulphur dioxide concentration related to the six other variables?" In the first instance at least, this question suggests the application of multiple linear regression, with sulphur dioxide concentration as the response variable and the remaining six variables being the independent or explanatory variables (the latter is a more acceptable label because the "independent" variables are rarely independent of one another). But in the model underlying multiple regression, only the response is considered to be a random variable; the explanatory variables are strictly assumed to be fixed, not random, variables. In practise, of course, this is rarely the case, and so the results from a multiple regression analysis need to be interpreted as being conditional on the observed values of the explanatory variables. So when answering the question of most interest about these data, they should not really be considered multivariate–there is only a single random variable involved–a more suitable label is *multivariable* (we know this sounds pedantic,

but we are statisticians after all). In this book, we shall say only a little about
the multiple linear model for multivariable data in Chapter 8. but essentially
only to enable such regression models to be introduced for situations where
there is a multivariate response; for example, in the case of *repeated-measures
data* and *longitudinal data*.

The four data sets above have not exhausted either the questions that
multivariate data may have been collected to answer or the methods of multivariate analysis that have been developed to answer them, as we shall see as
we progress through the book.

1.5 Covariances, correlations, and distances

The main reason why we should analyse a multivariate data set using multivariate methods rather than looking at each variable separately using one or
another familiar univariate method is that any structure or pattern in the data
is as likely to be implied either by "relationships" between the variables or by
the relative "closeness" of different units as by their different variable values;
in some cases perhaps by both. In the first case, any structure or pattern un-
covered will be such that it "links" together the columns of the data matrix,
X, in some way, and in the second case a possible structure that might be
discovered is that involving interesting subsets of the units. The question now
arises as to how we quantify the relationships between the variables and how
we measure the distances between different units. This question is answered
in the subsections that follow.

1.5.1 Covariances

The *covariance* of two random variables is a measure of their *linear* dependence. The population (theoretical) covariance of two random variables, X_i
and X_j, is defined by

$$\mathsf{Cov}(X_i, X_j) = \mathsf{E}(X_i - \mu_i)(X_j - \mu_j),$$

where $\mu_i = \mathsf{E}(X_i)$ and $\mu_j = \mathsf{E}(X_j)$; E denotes expectation.

If $i = j$, we note that the covariance of the variable with itself is simply its
variance, and therefore there is no need to define variances and covariances
independently in the multivariate case. If X_i and X_j are independent of each
other, their covariance is necessarily equal to zero, but the converse is not
true. The covariance of X_i and X_j is usually denoted by σ_{ij}. The variance of
variable X_i is $\sigma_i^2 = \mathsf{E}\left((X_i - \mu_i)^2\right)$. Larger values of the covariance imply a
greater degree of linear dependence between two variables.

In a multivariate data set with q observed variables, there are q variances
and $q(q - 1)/2$ covariances. These quantities can be conveniently arranged in
a $q \times q$ symmetric matrix, $\mathbf{\Sigma}$, where

$$\Sigma = \begin{pmatrix} \sigma_1^2 & \sigma_{12} & \cdots & \sigma_{1q} \\ \sigma_{21} & \sigma_2^2 & \cdots & \sigma_{2q} \\ \vdots & \vdots & \ddots & \vdots \\ \sigma_{q1} & \sigma_{q2} & \cdots & \sigma_q^2 \end{pmatrix}.$$

Note that $\sigma_{ij} = \sigma_{ji}$. This matrix is generally known as the *variance-covariance matrix* or simply the *covariance matrix* of the data.

For a set of multivariate observations, perhaps sampled from some population, the matrix Σ is estimated by

$$\mathbf{S} = \frac{1}{n-1} \sum_{i=1}^{n} (\mathbf{x}_i - \bar{\mathbf{x}})(\mathbf{x}_i - \bar{\mathbf{x}})^\top,$$

where $\mathbf{x}_i^\top = (x_{i1}, x_{i2}, \ldots, x_{iq})$ is the vector of (numeric) observations for the ith individual and $\bar{\mathbf{x}} = n^{-1} \sum_{i=1}^{n} \mathbf{x}_i$ is the mean vector of the observations. The diagonal of $\mathbf{S}$ contains the sample variances of each variable, which we shall denote as s_i^2.

The covariance matrix for the data in Table 1.2 can be obtained using the var() function in R; however, we have to "remove" the categorical variable gender from the measure data frame by subsetting on the numerical variables first:

```
R> cov(measure[, c("chest", "waist", "hips")])
```

```
      chest  waist   hips
chest 6.632  6.368 3.000
waist 6.368 12.526 3.579
hips  3.000  3.579 5.945
```

If we require the separate covariance matrices of men and women, we can use

```
R> cov(subset(measure, gender == "female")[,
+              c("chest", "waist", "hips")])
```

```
      chest waist  hips
chest 2.278 2.167 1.556
waist 2.167 2.989 2.756
hips  1.556 2.756 3.067
```

```
R> cov(subset(measure, gender == "male")[,
+              c("chest", "waist", "hips")])
```

```
       chest  waist  hips
chest 6.7222 0.9444 3.944
waist 0.9444 2.1000 3.078
hips  3.9444 3.0778 9.344
```

where the subset() returns all observations corresponding to females (first statement) or males (second statement).

1.5.2 Correlations

The covariance is often difficult to interpret because it depends on the scales on which the two variables are measured; consequently, it is often standardised by dividing by the product of the standard deviations of the two variables to give a quantity called the *correlation coefficient*, ρ_{ij}, where

$$\rho_{ij} = \frac{\sigma_{ij}}{\sigma_i \sigma_j},$$

where $\sigma_i = \sqrt{\sigma_i^2}$.

The advantage of the correlation is that it is independent of the scales of the two variables. The correlation coefficient lies between -1 and $+1$ and gives a measure of the *linear* relationship of the variables X_i and X_j. It is positive if high values of X_i are associated with high values of X_j and negative if high values of X_i are associated with low values of X_j. If the relationship between two variables is non-linear, their correlation coefficient can be misleading.

With q variables there are $q(q-1)/2$ distinct correlations, which may be arranged in a $q \times q$ correlation matrix the diagonal elements of which are unity. For observed data, the correlation matrix contains the usual estimates of the ρs, namely Pearson's correlation coefficient, and is generally denoted by $\mathbf{R}$. The matrix may be written in terms of the sample covariance matrix $\mathbf{S}$

$$\mathbf{R} = \mathbf{D}^{-1/2}\mathbf{S}\mathbf{D}^{-1/2},$$

where $\mathbf{D}^{-1/2} = \text{diag}(1/s_1, \ldots, 1/s_q)$ and $s_i = \sqrt{s_i^2}$ is the sample standard deviation of variable i. (In most situations considered in this book, we will be dealing with covariance and correlation matrices of full rank, q, so that both matrices will be *non-singular*, that is, invertible, to give matrices $\mathbf{S}^{-1}$ or $\mathbf{R}^{-1}$.)

The sample correlation matrix for the three variables in Table 1.1 is obtained by using the function `cor()` in R:

```
R> cor(measure[, c("chest", "waist", "hips")])

       chest  waist   hips
chest 1.0000 0.6987 0.4778
waist 0.6987 1.0000 0.4147
hips  0.4778 0.4147 1.0000
```

1.5.3 Distances

For some multivariate techniques such as *multidimensional scaling* (see Chapter 4) and *cluster analysis* (see Chapter 6), the concept of *distance* between the units in the data is often of considerable interest and importance. So, given the variable values for two units, say unit i and unit j, what serves as a measure of distance between them? The most common measure used is *Euclidean distance*, which is defined as

$$d_{ij} = \sqrt{\sum_{k=1}^{q}(x_{ik} - x_{jk})^2},$$

where x_{ik} and $x_{jk}, k = 1, \ldots, q$ are the variable values for units i and j, respectively. Euclidean distance can be calculated using the dist() function in R.

When the variables in a multivariate data set are on different scales, it makes more sense to calculate the distances *after* some form of standardisation. Here we shall illustrate this on the body measurement data and divide each variable by its standard deviation using the function scale() before applying the dist() function–the necessary R code and output are

```
R> dist(scale(measure[, c("chest", "waist", "hips")],
+        center = FALSE))
```

```
       1    2    3    4    5    6    7    8    9    10   11
2   0.17
3   0.15 0.08
4   0.22 0.07 0.14
5   0.11 0.15 0.09 0.22
6   0.29 0.16 0.16 0.19 0.21
7   0.32 0.16 0.20 0.13 0.28 0.14
8   0.23 0.11 0.11 0.12 0.19 0.16 0.13
9   0.21 0.10 0.06 0.16 0.12 0.11 0.17 0.09
10  0.27 0.12 0.13 0.14 0.20 0.06 0.09 0.11 0.09
11  0.23 0.28 0.22 0.33 0.19 0.34 0.38 0.25 0.24 0.32
12  0.22 0.24 0.18 0.28 0.18 0.30 0.32 0.20 0.20 0.28 0.06
```

. . .

(Note that only the distances for the first 12 observations are shown in the output.)

1.6 The multivariate normal density function

Just as the normal distribution dominates univariate techniques, the *multivariate normal distribution* plays an important role in *some* multivariate procedures, although as mentioned earlier many multivariate analyses are carried out in the spirit of data exploration where questions of statistical significance are of relatively minor importance or of no importance at all. Nevertheless, researchers dealing with the complexities of multivariate data may, on occasion, need to know a little about the multivariate density function and in particular how to assess whether or not a set of multivariate data can be assumed to have this density function. So we will define the multivariate normal density and describe some of its properties.

For a vector of q variables, $\mathbf{x}^\top = (x_1, x_2, \ldots, x_q)$, the multivariate normal density function takes the form

$$f(\mathbf{x}; \boldsymbol{\mu}, \boldsymbol{\Sigma}) = (2\pi)^{-q/2} \det(\boldsymbol{\Sigma})^{-1/2} \exp\left\{-\frac{1}{2}(\mathbf{x} - \boldsymbol{\mu})^\top \boldsymbol{\Sigma}^{-1}(\mathbf{x} - \boldsymbol{\mu})\right\},$$

where $\boldsymbol{\Sigma}$ is the population covariance matrix of the variables and $\boldsymbol{\mu}$ is the vector of population mean values of the variables. The simplest example of the *multivariate normal density function* is the bivariate normal density with $q = 2$; this can be written explicitly as

$$f((x_1, x_2); (\mu_1, \mu_2), \sigma_1, \sigma_2, \rho) =$$
$$\left(2\pi\sigma_1\sigma_2(1 - \rho^2)\right)^{-1/2} \exp\left\{-\frac{1}{2(1 - \rho^2)} \times\right.$$
$$\left.\left(\left(\frac{x_1 - \mu_1}{\sigma_1}\right)^2 - 2\rho\frac{x_1 - \mu_1}{\sigma_1}\frac{x_2 - \mu_2}{\sigma_2} + \left(\frac{x_2 - \mu_2}{\sigma_2}\right)^2\right)\right\},$$

where μ_1 and μ_2 are the population means of the two variables, σ_1^2 and σ_2^2 are the population variances, and ρ is the population correlation between the two variables X_1 and X_2. Figure 1.1 shows an example of a bivariate normal density function with both means equal to zero, both variances equal to one, and correlation equal to 0.5.

The population mean vector and the population covariance matrix of a multivariate density function are estimated from a sample of multivariate observations as described in the previous subsections.

One property of a multivariate normal density function that is worth mentioning here is that *linear combinations* of the variables (i.e., $y = a_1X_1 + a_2X_2 + \cdots + a_qX_q$, where $a_1, a_2, \ldots, a_q$ is a set of scalars) are themselves normally distributed with mean $\mathbf{a}^\top \boldsymbol{\mu}$ and variance $\mathbf{a}^\top \boldsymbol{\Sigma} \mathbf{a}$, where $\mathbf{a}^\top = (a_1, a_2, \ldots, a_q)$. Linear combinations of variables will be of importance in later chapters, particularly in Chapter 3.

For many multivariate methods to be described in later chapters, the assumption of multivariate normality is not critical to the results of the analysis, but there may be occasions when testing for multivariate normality may be of interest. A start can be made perhaps by assessing each variable separately for univariate normality using a *probability plot*. Such plots are commonly applied in univariate analysis and involve ordering the observations and then plotting them against the appropriate values of an assumed cumulative distribution function. There are two basic types of plots for comparing two probability distributions, the *probability-probability plot* and the *quantile-quantile plot*. The diagram in Figure 1.2 may be used for describing each type.

A plot of points whose coordinates are the cumulative probabilities $p_1(q)$ and $p_2(q)$ for different values of q with

$$p_1(q) = \mathsf{P}(X_1 \leq q),$$
$$p_2(q) = \mathsf{P}(X_2 \leq q),$$

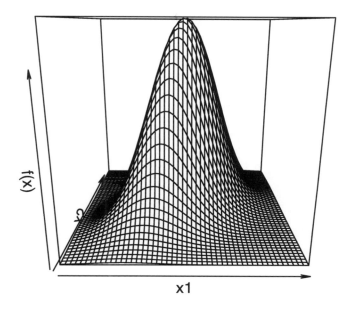

Fig. 1.1. Bivariate normal density function with correlation $\rho = 0.5$.

for random variables X_1 and X_2 is a probability-probability plot, while a plot of the points whose coordinates are the quantiles $(q_1(p), q_2(p))$ for different values of p with

$$q_1(p) = p_1^{-1}(p),$$
$$q_2(p) = p_2^{-1}(p),$$

is a quantile-quantile plot. For example, a quantile-quantile plot for investigating the assumption that a set of data is from a normal distribution would involve plotting the ordered sample values of variable 1 (i.e., $x_{(1)1}, x_{(2)1}, \ldots, x_{(n)1}$) against the quantiles of a standard normal distribution, $\Phi^{-1}(p(i))$, where usually

$$p_i = \frac{i - \frac{1}{2}}{n} \quad \Phi(x) = \int_{-\infty}^{x} \frac{1}{\sqrt{2\pi}} e^{-\frac{1}{2}u^2} \, du.$$

This is known as a *normal probability plot*.

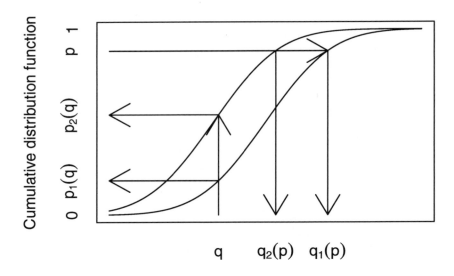

Fig. 1.2. Cumulative distribution functions and quantiles.

For multivariate data, normal probability plots may be used to examine each variable separately, although marginal normality does not necessarily imply that the variables follow a multivariate normal distribution. Alternatively (or additionally), each multivariate observation might be converted to a single number in some way before plotting. For example, in the specific case of assessing a data set for multivariate normality, each q-dimensional observation, $\mathbf{x}_i$, could be converted into a *generalised distance*, d_i^2, giving a measure of the distance of the particular observation from the mean vector of the complete sample, $\bar{\mathbf{x}}$; d_i^2 is calculated as

$$d_i^2 = (\mathbf{x}_i - \bar{\mathbf{x}})^\top \mathbf{S}^{-1}(\mathbf{x}_i - \bar{\mathbf{x}}),$$

where $\mathbf{S}$ is the sample covariance matrix. This distance measure takes into account the different variances of the variables and the covariances of pairs of variables. If the observations do arise from a multivariate normal distribution, then these distances have approximately a *chi-squared distribution* with q degrees of freedom, also denoted by the symbol χ_q^2. So plotting the ordered distances against the corresponding quantiles of the appropriate chi-square distribution should lead to a straight line through the origin.

We will now assess the body measurements data in Table 1.2 for normality, although because there are only 20 observations in the sample there is really too little information to come to any convincing conclusion. Figure 1.3 shows separate probability plots for each measurement; there appears to be no evidence of any departures from linearity. The chi-square plot of the 20 generalised distances in Figure 1.4 does seem to deviate a little from linearity, but with so few observations it is hard to be certain. The plot is set up as follows. We first extract the relevant data

```
R> x <- measure[, c("chest", "waist", "hips")]
```

and estimate the means of all three variables (i.e., for each column of the data) and the covariance matrix

```
R> cm <- colMeans(x)
R> S <- cov(x)
```

The differences d_i have to be computed for all units in our data, so we iterate over the rows of x using the apply() function with argument MARGIN = 1 and, for each row, compute the distance d_i:

```
R> d <- apply(x, MARGIN = 1, function(x)
+             t(x - cm) %*% solve(S) %*% (x - cm))
```

The sorted distances can now be plotted against the appropriate quantiles of the χ_3^2 distribution obtained from qchisq(); see Figure 1.4.

```
R> qqnorm(measure[,"chest"], main = "chest"); qqline(measure[,"chest"])
R> qqnorm(measure[,"waist"], main = "waist"); qqline(measure[,"waist"])
R> qqnorm(measure[,"hips"], main = "hips"); qqline(measure[,"hips"])
```

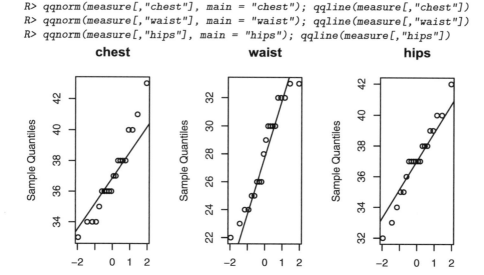

Fig. 1.3. Normal probability plots of chest, waist, and hip measurements.

```
R> plot(qchisq((1:nrow(x) - 1/2) / nrow(x), df = 3), sort(d),
+        xlab = expression(paste(chi[3]^2, " Quantile")),
+        ylab = "Ordered distances")
R> abline(a = 0, b = 1)
```

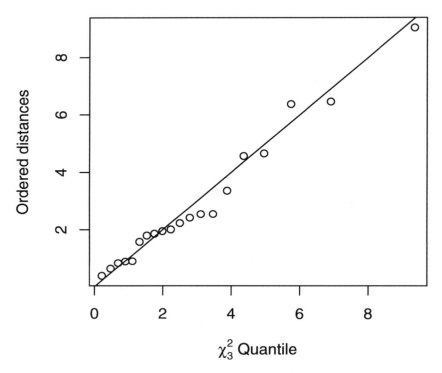

Fig. 1.4. Chi-square plot of generalised distances for body measurements data.

We will now look at using the chi-square plot on a set of data introduced early in the chapter, namely the air pollution in US cities (see Table 1.5). The probability plots for each separate variable are shown in Figure 1.5. Here, we also iterate over all variables, this time using a special function, sapply(), that loops over the variable names:

```
R> layout(matrix(1:8, nc = 2))
R> sapply(colnames(USairpollution), function(x) {
+        qqnorm(USairpollution[[x]], main = x)
+        qqline(USairpollution[[x]])
+ })
```

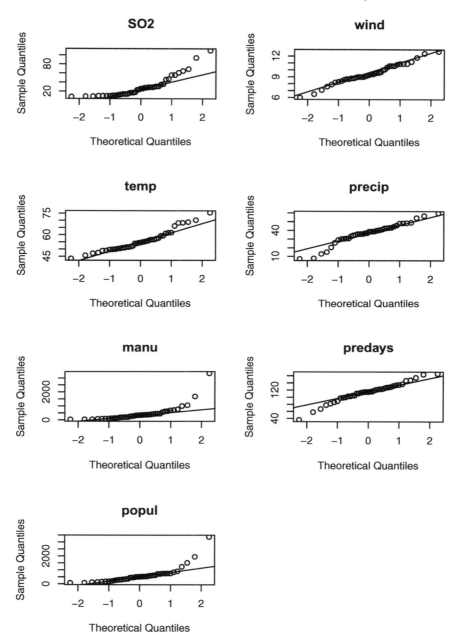

Fig. 1.5. Normal probability plots for USairpollution data.

The resulting seven plots are arranged on one page by a call to the `layout` matrix; see Figure 1.5. The plots for SO_2 concentration and precipitation both deviate considerably from linearity, and the plots for manufacturing and population show evidence of a number of outliers. But of more importance is the chi-square plot for the data, which is given in Figure 1.6; the R code is identical to the code used to produce the chi-square plot for the body measurement data. In addition, the two most extreme points in the plot have been labelled with the city names to which they correspond using `text()`.

```
R> x <- USairpollution
R> cm <- colMeans(x)
R> S <- cov(x)
R> d <- apply(x, 1, function(x) t(x - cm) %*% solve(S) %*% (x - cm))
R> plot(qc <- qchisq((1:nrow(x) - 1/2) / nrow(x), df = 6),
+        sd <- sort(d),
+        xlab = expression(paste(chi[6]^2, " Quantile")),
+        ylab = "Ordered distances", xlim = range(qc) * c(1, 1.1))
R> oups <- which(rank(abs(qc - sd), ties = "random") > nrow(x) - 3)
R> text(qc[oups], sd[oups] - 1.5, names(oups))
R> abline(a = 0, b = 1)
```

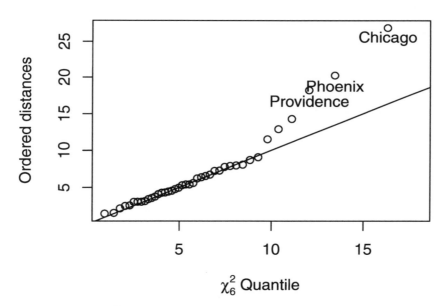

Fig. 1.6. χ^2 plot of generalised distances for `USairpollution` data.

This example illustrates that the chi-square plot might also be useful for detecting possible outliers in multivariate data, where informally outliers are "abnormal" in the sense of deviating from the natural data variability. Outlier identification is important in many applications of multivariate analysis either because there is some specific interest in finding anomalous observations or as a pre-processing task before the application of some multivariate method in order to preserve the results from possible misleading effects produced by these observations. A number of methods for identifying multivariate outliers have been suggested–see, for example, Rocke and Woodruff (1996) and Becker and Gather (2001)–and in Chapter 2 we will see how a number of the graphical methods described there can also be helpful for outlier detection.

1.7 Summary

The majority of data collected in all scientific disciplines are multivariate. To fully understand most such data sets, the variables need to be analysed simultaneously. The rest of this text is concerned with methods that have been developed to make this possible, some with the aim of discovering any patterns or structure in the data that may have important implications for future studies and some with the aim of drawing inferences about the data assuming they are sampled from a population with some particular probability density function, usually the multivariate normal.

1.8 Exercises

Ex. 1.1 Find the correlation matrix and covariance matrix of the data in Table 1.1.

Ex. 1.2 Fill in the missing values in Table 1.1 with appropriate mean values, and recalculate the correlation matrix of the data.

Ex. 1.3 Examine both the normal probability plots of each variable in the archaeology data in Table 1.3 and the chi-square plot of the data. Do the plots suggest anything unusual about the data?

Ex. 1.4 Convert the covariance matrix given below into the corresponding correlation matrix.

$$\begin{pmatrix} 3.8778 & 2.8110 & 3.1480 & 3.5062 \\ 2.8110 & 2.1210 & 2.2669 & 2.5690 \\ 3.1480 & 2.2669 & 2.6550 & 2.8341 \\ 3.5062 & 2.5690 & 2.8341 & 3.2352 \end{pmatrix}.$$

Ex. 1.5 For the small set of (10×5) multivariate data given below, find the (10×10) Euclidean distance matrix for the rows of the matrix. An alternative to Euclidean distance that might be used in some cases is what is known as *city block distance* (think New York). Write some R code to calculate the city block distance matrix for the data.

$$
\begin{pmatrix}
3 & 6 & 4 & 0 & 7 \\
4 & 2 & 7 & 4 & 6 \\
4 & 0 & 3 & 1 & 5 \\
6 & 2 & 6 & 1 & 1 \\
1 & 6 & 2 & 1 & 4 \\
5 & 1 & 2 & 0 & 2 \\
1 & 1 & 2 & 6 & 1 \\
1 & 1 & 5 & 4 & 4 \\
7 & 0 & 1 & 3 & 3 \\
3 & 3 & 0 & 5 & 1
\end{pmatrix} .
$$

2

Looking at Multivariate Data: Visualisation

2.1 Introduction

According to Chambers, Cleveland, Kleiner, and Tukey (1983), "there is no statistical tool that is as powerful as a well-chosen graph". Certainly graphical presentation has a number of advantages over tabular displays of numerical results, not least in creating interest and attracting the attention of the viewer. But just what is a graphical display? A concise description is given by Tufte (1983):

> Data graphics visually display measured quantities by means of the combined use of points, lines, a coordinate system, numbers, symbols, words, shading and color.

Graphs are very popular; it has been estimated that between 900 billion (9×10^{11}) and 2 trillion (2×10^{12}) images of statistical graphics are printed each year. Perhaps one of the main reasons for such popularity is that graphical presentation of data often provides the vehicle for discovering the unexpected; the human visual system is very powerful in detecting patterns, although the following caveat from the late Carl Sagan (in his book *Contact*) should be kept in mind:

> Humans are good at discerning subtle patterns that are really there, but equally so at imagining them when they are altogether absent.

Some of the advantages of graphical methods have been listed by Schmid (1954):

- In comparison with other types of presentation, well-designed charts are more effective in creating interest and in appealing to the attention of the reader.
- Visual relationships as portrayed by charts and graphs are more easily grasped and more easily remembered.
- The use of charts and graphs saves time since the essential meaning of large measures of statistical data can be visualised at a glance.

- Charts and graphs provide a comprehensive picture of a problem that makes for a more complete and better balanced understanding than could be derived from tabular or textual forms of presentation.
- Charts and graphs can bring out hidden facts and relationships and can stimulate, as well as aid, analytical thinking and investigation.

Schmid's last point is reiterated by the legendary John Tukey in his observation that "the greatest value of a picture is when it forces us to notice what we never expected to see".

The prime objective of a graphical display is to communicate to ourselves and others, and the graphic design must do everything it can to help people understand. And unless graphics are relatively simple, they are unlikely to survive the first glance. There are perhaps four goals for graphical displays of data:

- To provide an overview;
- To tell a story;
- To suggest hypotheses;
- To criticise a model.

In this chapter, we will be largely concerned with graphics for multivariate data that address one or another of the first three bulleted points above. Graphics that help in checking model assumptions will be considered in Chapter 8.

During the last two decades, a wide variety of new methods for displaying data graphically have been developed. These will hunt for special effects in data, indicate outliers, identify patterns, diagnose models, and generally search for novel and perhaps unexpected phenomena. Graphical displays should aim to tell a story about the data and to reduce the cognitive effort required to make comparisons. Large numbers of graphs might be required to achieve these goals, and computers are generally needed to supply them for the same reasons that they are used for numerical analyses, namely that they are fast and accurate.

So, because the machine is doing the work, the question is no longer "shall we plot?" but rather "what shall we plot?" There are many exciting possibilities, including interactive and dynamic graphics on a computer screen (see Cook and Swayne 2007), but graphical exploration of data usually begins at least with some simpler *static* graphics. The starting graphic for multivariate data is often the ubiquitous *scatterplot*, and this is the subject of the next section.

2.2 The scatterplot

The simple xy scatterplot has been in use since at least the 18th century and has many virtues–indeed, according to Tufte (1983):

The relational graphic–in its barest form the scatterplot and its variants–is the greatest of all graphical designs. It links at least two variables, encouraging and even imploring the viewer to assess the possible causal relationship between the plotted variables. It confronts causal theories that x causes y with empirical evidence as to the actual relationship between x and y.

The scatterplot is the standard for representing continuous *bivariate data* but, as we shall see later in this chapter, it can be enhanced in a variety of ways to accommodate information about other variables.

To illustrate the use of the scatterplot and a number of other techniques to be discussed, we shall use the air pollution in US cities data introduced in the previous chapter (see Table 1.5).

Let's begin our examination of the air pollution data by taking a look at a basic scatterplot of the two variables `manu` and `popul`. For later use, we first set up two character variables that contain the labels to be printed on the two axes:

```
R> mlab <- "Manufacturing enterprises with 20 or more workers"
R> plab <- "Population size (1970 census) in thousands"
```

The `plot()` function takes the data, here as the data frame `USairpollution`, along with a "`formula`" describing the variables to be plotted; the part left of the tilde defines the variable to be associated with the ordinate, the part right of the tilde is the variable that goes with the abscissa:

```
R> plot(popul ~ manu, data = USairpollution,
+        xlab = mlab, ylab = plab)
```

The resulting scatterplot is shown in Figure 2.2. The plot clearly uncovers the presence of one or more cities that are some way from the remainder, but before commenting on these possible *outliers* we will construct the scatterplot again but now show how to include the marginal distributions of `manu` and `popul` in two different ways. Plotting marginal and joint distributions together is usually good data analysis practise. In Figure 2.2, the marginal distributions are shown as rug plots on each axis (produced by `rug()`), and in Figure 2.3 the marginal distribution of `manu` is given as a histogram and that of `popul` as a boxplot. And also in Figure 2.3 the points are labelled by an abbreviated form of the corresponding city name.

The necessary R code for Figure 2.3 starts with dividing the device into three plotting areas by means of the `layout()` function. The first plot basically resembles the `plot()` command from Figure 2.1, but instead of points the abbreviated name of the city is used as the plotting symbol. Finally, the `hist()` and `boxplots()` commands are used to depict the marginal distributions. The `with()` command is very useful when one wants to avoid explicitly extracting variables from data frames. The command of interest, here the calls to `hist()` and `boxplot()`, is evaluated "inside" the data frame, here `USairpollution` (i.e., variable names are resolved within this data frame first).

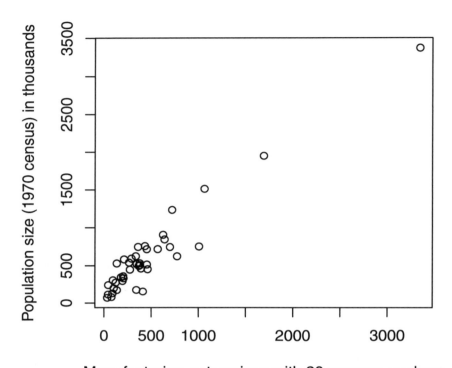

Fig. 2.1. Scatterplot of `manu` and `popul`.

From this series of plots, we can see that the outlying points show themselves in both the scatterplot of the variables *and* in each marginal distribution. The most extreme outlier corresponds to Chicago, and other slightly less extreme outliers correspond to Philadelphia and Detroit. Each of these cities has a considerably larger population than other cities and also many more manufacturing enterprises with more than 20 workers.

2.2.1 The bivariate boxplot

In Figure 2.3, identifying Chicago, Philadelphia, and Detroit as outliers is unlikely to invoke much argument, but what about Houston and Cleveland? In many cases, it might be helpful to have a more formal and objective method for labelling observations as outliers, and such a method is provided by the *bivariate boxplot*, which is a two-dimensional analogue of the boxplot for univariate data proposed by Goldberg and Iglewicz (1992). This type of graphic

```
R> plot(popul ~ manu, data = USairpollution,
+        xlab = mlab, ylab = plab)
R> rug(USairpollution$manu, side = 1)
R> rug(USairpollution$popul, side = 2)
```

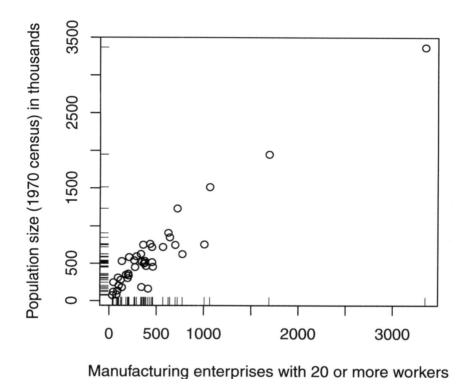

Fig. 2.2. Scatterplot of manu and popul that shows the marginal distribution in each variable as a rug plot.

may be useful in indicating the distributional properties of the data and in identifying possible outliers. The bivariate boxplot is based on calculating "robust" measures of location, scale, and correlation; it consists essentially of a pair of concentric ellipses, one of which (the "hinge") includes 50% of the data and the other (called the "fence") of which delineates potentially troublesome outliers. In addition, resistant regression lines of both y on x and x on y are shown, with their intersection showing the bivariate location estimator. The acute angle between the regression lines will be small for a large absolute value of correlations and large for a small one. (Using robust measures of location, scale, etc., helps to prevent the possible "masking" of multivariate outliers if

```
R> layout(matrix(c(2, 0, 1, 3), nrow = 2, byrow = TRUE),
+          widths = c(2, 1), heights = c(1, 2), respect = TRUE)
R> xlim <- with(USairpollution, range(manu)) * 1.1
R> plot(popul ~ manu, data = USairpollution, cex.lab = 0.9,
+        xlab = mlab, ylab = plab, type = "n", xlim = xlim)
R> with(USairpollution, text(manu, popul, cex = 0.6,
+        labels = abbreviate(row.names(USairpollution))))
R> with(USairpollution, hist(manu, main = "", xlim = xlim))
R> with(USairpollution, boxplot(popul))
```

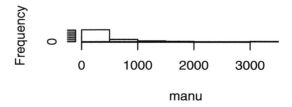

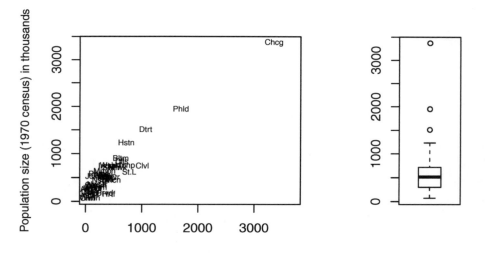

Fig. 2.3. Scatterplot of manu and popul that shows the marginal distributions by histogram and boxplot.

the usual measures are employed when these may be distorted by the presence of the outliers in the data.) Full details of the construction are given in Goldberg and Iglewicz (1992). The scatterplot of manu and popul including the bivariate boxplot is shown in Figure 2.4. Figure 2.4 clearly tells us that Chicago, Philadelphia, Detroit, and Cleveland should be regarded as outliers but not Houston, because it is on the "fence" rather than outside the "fence".

```
R> outcity <- match(lab <- c("Chicago", "Detroit",
+      "Cleveland", "Philadelphia"), rownames(USairpollution))
R> x <- USairpollution[, c("manu", "popul")]
R> bvbox(x, mtitle = "", xlab = mlab, ylab = plab)
R> text(x$manu[outcity], x$popul[outcity], labels = lab,
+      cex = 0.7, pos = c(2, 2, 4, 2, 2))
```

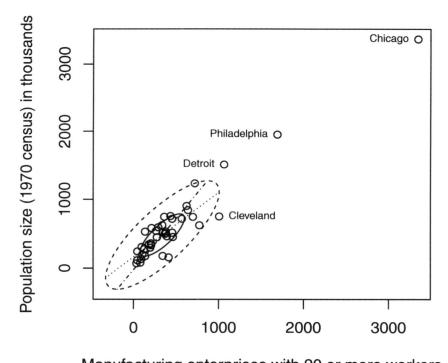

Fig. 2.4. Scatterplot of manu and popul showing the bivariate boxplot of the data.

Suppose now that we are interested in calculating the correlation between manu and popul. Researchers often calculate the correlation between two variables *without* first looking at the scatterplot of the two variables. But scatterplots should *always* be consulted when calculating correlation coefficients because the presence of outliers can on occasion considerably distort the value of a correlation coefficient, and as we have seen above, a scatterplot may help to identify the offending observations particularly if used in conjunction with a bivariate boxplot. The observations identified as outliers may then be excluded from the calculation of the correlation coefficient. With the help of the bivariate boxplot in Figure 2.4, we have identified Chicago, Philadelphia, Detroit, and Cleveland as outliers in the scatterplot of `manu` and `popul`. The R code for finding the two correlations is

```
R> with(USairpollution, cor(manu, popul))
```

```
[1] 0.9553
```

```
R> outcity <- match(c("Chicago", "Detroit",
+                      "Cleveland", "Philadelphia"),
+                      rownames(USairpollution))
R> with(USairpollution, cor(manu[-outcity], popul[-outcity]))
```

```
[1] 0.7956
```

The `match()` function identifies rows of the data frame `USairpollution` corresponding to the cities of interest, and the subset starting with a minus sign removes these units before the correlation is computed. Calculation of the correlation coefficient between the two variables using all the data gives a value of 0.96, which reduces to a value of 0.8 after excluding the four outliers–a not inconsiderable reduction.

2.2.2 The convex hull of bivariate data

An alternative approach to using the scatterplot combined with the bivariate boxplot to deal with the possible problem of calculating correlation coefficients without the distortion often caused by outliers in the data is *convex hull trimming*, which allows *robust estimation* of the correlation. The convex hull of a set of bivariate observations consists of the vertices of the smallest convex polyhedron in variable space within which or on which all data points lie. Removal of the points lying on the convex hull can eliminate isolated outliers without disturbing the general shape of the bivariate distribution. A robust estimate of the correlation coefficient results from using the remaining observations. Let's see how the convex hull approach works with our `manu` and `popul` scatterplot. We first find the convex hull of the data (i.e., the observations defining the convex hull) using the following R code:

```
R> (hull <- with(USairpollution, chull(manu, popul)))
```

```
[1]  9 15 41  6  2 18 16 14  7
```

```
R> with(USairpollution,
+        plot(manu, popul, pch = 1, xlab = mlab, ylab = plab))
R> with(USairpollution,
+        polygon(manu[hull], popul[hull], density = 15, angle = 30))
```

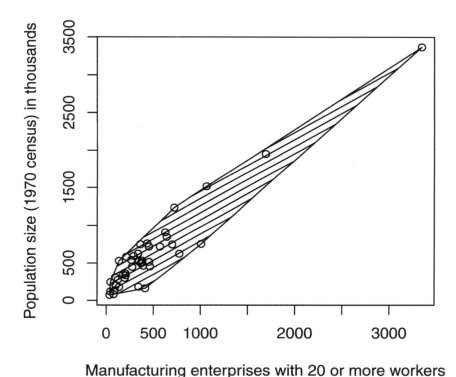

Fig. 2.5. Scatterplot of manu against popul showing the convex hull of the data.

Now we can show this convex hull on a scatterplot of the variables using the code attached to the resulting Figure 2.5.

To calculate the correlation coefficient after removal of the points defining the convex hull requires the code

```
R> with(USairpollution, cor(manu[-hull],popul[-hull]))
```

```
[1] 0.9225
```

The resulting value of the correlation is now 0.923 and thus is higher compared with the correlation estimated after removal of the outliers identified by using the bivariate boxplot, namely Chicago, Philadelphia, Detroit, and Cleveland.

2.2.3 The chi-plot

Although the scatterplot is a primary data-analytic tool for assessing the relationship between a pair of continuous variables, it is often difficult to judge whether or not the variables are independent–a random scatter of points is hard for the human eye to judge. Consequently it is sometimes helpful to augment the scatterplot with an auxiliary display in which independence is itself manifested in a characteristic manner. The *chi-plot* suggested by Fisher and Switzer (1985, 2001) is designed to address the problem. Under independence, the joint distribution of two random variables X_1 and X_2 can be computed from the product of the marginal distributions. The chi-plot transforms the measurements $(x_{11}, \ldots, x_{n1})$ and $(x_{12}, \ldots, x_{n2})$ into values $(\chi_1, \ldots, \chi_n)$ and $(\lambda_1, \ldots, \lambda_n)$, which, plotted in a scatterplot, can be used to detect deviations from independence. The χ_i values are, basically, the root of the χ^2 statistics obtained from the 2×2 tables that are obtained when dichotomising the data for each unit i into the groups satisfying $x_{\cdot 1} \leq x_{i1}$ and $x_{\cdot 2} \leq x_{i2}$. Under independence, these values are asymptotically normal with mean zero; i.e., the χ_i values should show a non-systematic random fluctuation around zero. The λ_i values measure the distance of unit i from the "center" of the bivariate distribution. An R function for producing chi-plots is `chiplot()`. To illustrate the chi-plot, we shall apply it to the `manu` and `popul` variables of the air pollution data using the code

```
R> with(USairpollution, plot(manu, popul,
+                            xlab = mlab, ylab = plab,
+                            cex.lab = 0.9))
R> with(USairpollution, chiplot(manu, popul))
```

The result is Figure 2.6, which shows the scatterplot of `manu` plotted against `popul` alongside the corresponding chi-plot. Departure from independence is indicated in the latter by a lack of points in the horizontal band indicated on the plot. Here there is a very clear departure since there are very few of the observations in this region.

2.3 The bubble and other glyph plots

The basic scatterplot can only display two variables. But there have been a number of suggestions as to how extra variables may be included on a scatterplot. Perhaps the simplest is the so-called *bubble plot*, in which three variables are displayed; two are used to form the scatterplot itself, and then the values of the third variable are represented by circles with radii proportional to these values and centred on the appropriate point in the scatterplot. Let's begin by taking a look at the bubble plot of `temp`, `wind`, and `SO2` that is given in Figure 2.7. The plot seems to suggest that cities with moderate annual temperatures and moderate annual wind speeds tend to suffer the greatest air

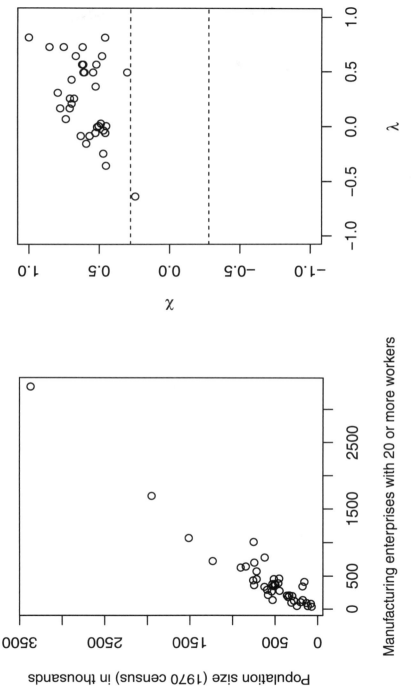

Fig. 2.6. Chi-plot for manu and popul showing a clear deviation from independence.

pollution, but this is unlikely to be the whole story because none of the other variables in the data set are used in constructing Figure 2.7. We could try to include all variables on the basic `temp` and `wind` scatterplot by replacing the circles with five-sided "stars", with the lengths of each side representing each of the remaining five variables. Such a plot is shown in Figure 2.8, but it fails to communicate much, if any, useful information about the data.

```
R> ylim <- with(USairpollution, range(wind)) * c(0.95, 1)
R> plot(wind ~ temp, data = USairpollution,
+        xlab = "Average annual temperature (Fahrenheit)",
+        ylab = "Average annual wind speed (m.p.h.)", pch = 10,
+        ylim = ylim)
R> with(USairpollution, symbols(temp, wind, circles = SO2,
+                               inches = 0.5, add = TRUE))
```

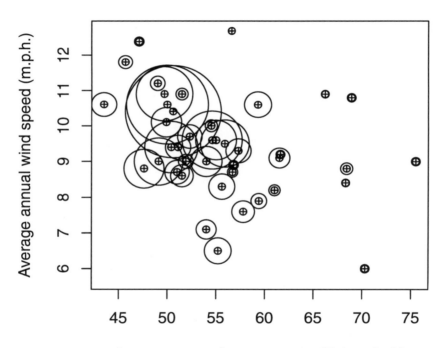

Fig. 2.7. Bubble plot of `temp`, `wind`, and SO2.

```
R> plot(wind ~ temp, data = USairpollution,
+       xlab = "Average annual temperature (Fahrenheit)",
+       ylab = "Average annual wind speed (m.p.h.)", pch = 10,
+       ylim = ylim)
R> with(USairpollution,
+       stars(USairpollution[,-c(2,5)], locations = cbind(temp, wind),
+             labels = NULL, add = TRUE, cex = 0.5))
```

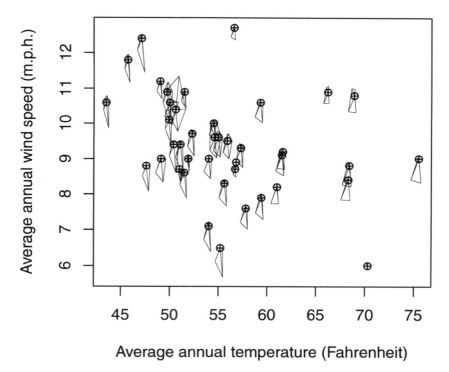

Fig. 2.8. Scatterplot of `temp` and `wind` showing five-sided stars representing the other variables.

In fact, both the bubble plot and "stars" plot are examples of *symbol* or *glyph plots*, in which data values control the symbol parameters. For example, a circle is a glyph where the values of one variable in a multivariate observation control the circle size. In Figure 2.8, the spatial positions of the cities in the scatterplot of `temp` and `wind` are combined with a star representation of the five other variables. An alternative is simply to represent the seven variables for each city by a seven-sided star and arrange the resulting stars in a rectangular array; the result is shown in Figure 2.9. We see that some stars, for example those for New Orleans, Miami, Jacksonville, and Atlanta, have similar shapes, with their higher average annual temperature being distinctive, but telling a story about the data with this display is difficult.

Stars, of course, are not the only symbols that could be used to represent data, and others have been suggested, with perhaps the most well known being the now infamous Chernoff's faces (see Chernoff 1973). But, on the whole, such graphics for displaying multivariate data have not proved themselves to be effective for the task and are now largely confined to the past history of multivariate graphics.

```
R> stars(USairpollution, cex = 0.55)
```

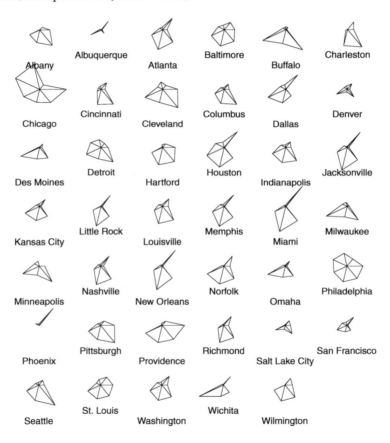

Fig. 2.9. Star plot of the air pollution data.

2.4 The scatterplot matrix

There are seven variables in the air pollution data, which between them generate 21 possible scatterplots. But just making the graphs without any coordination will often result in a confusing collection of graphs that are hard to integrate visually. Consequently, it is very important that the separate plots be presented in the best way to aid overall comprehension of the data. The *scatterplot matrix* is intended to accomplish this objective. A scatterplot matrix is nothing more than a square, symmetric grid of bivariate scatterplots. The grid has q rows and columns, each one corresponding to a different variable. Each of the grid's cells shows a scatterplot of two variables. Variable j is plotted against variable i in the ijth cell, and the same variables appear in cell ji, with the x- and y-axes of the scatterplots interchanged. The reason for including both the upper and lower triangles of the grid, despite the seeming redundancy, is that it enables a row and a column to be visually scanned to see one variable against all others, with the scales for the one variable lined up along the horizontal or the vertical. As a result, we can visually *link* features on one scatterplot with features on another, and this ability greatly increases the power of the graphic.

The scatterplot matrix for the air pollution data is shown in Figure 2.10. The plot was produced using the function pairs(), here with slightly enlarged dot symbols, using the arguments pch = "." and cex = 1.5.

The scatterplot matrix clearly shows the presence of possible outliers in many panels and the suggestion that the relationship between the two aspects of rainfall, namely precip, predays, and SO2 might be non-linear. Remembering that the *multivariable* aspect of these data, in which sulphur dioxide concentration is the response variable, with the remaining variables being explanatory, might be of interest, the scatterplot matrix may be made more helpful by including the linear fit of the two variables on each panel, and such a plot is shown in Figure 2.11. Here, the pairs() function was customised by a small function specified to the panel argument: in addition to plotting the x and y values, a regression line obtained via function lm() is added to each of the panels.

Now the scatterplot matrix reveals that there is a strong linear relationship between SO2 and manu and between SO2 and popul, but the (3, 4) panel shows that manu and popul are themselves very highly related and thus predictive of SO2 in the same way. Figure 2.11 also underlines that assuming a linear relationship between SO2 and precip and SO2 and predays, as might be the case if a multiple linear regression model is fitted to the data with SO2 as the dependent variable, is unlikely to fully capture the relationship between each pair of variables.

In the same way that the scatterplot should always be used alongside the numerical calculation of a correlation coefficient, so should the scatterplot matrix always be consulted when looking at the correlation matrix of a set of variables. The correlation matrix for the air pollution data is

```
R> pairs(USairpollution, pch = ".", cex = 1.5)
```

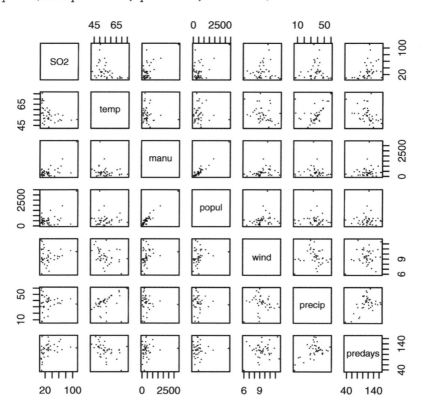

Fig. 2.10. Scatterplot matrix of the air pollution data.

```
R> round(cor(USairpollution), 4)
```

	SO2	temp	manu	popul	wind	precip	predays
SO2	1.0000	-0.4336	0.6448	0.4938	0.0947	0.0543	0.3696
temp	-0.4336	1.0000	-0.1900	-0.0627	-0.3497	0.3863	-0.4302
manu	0.6448	-0.1900	1.0000	0.9553	0.2379	-0.0324	0.1318
popul	0.4938	-0.0627	0.9553	1.0000	0.2126	-0.0261	0.0421
wind	0.0947	-0.3497	0.2379	0.2126	1.0000	-0.0130	0.1641
precip	0.0543	0.3863	-0.0324	-0.0261	-0.0130	1.0000	0.4961
predays	0.3696	-0.4302	0.1318	0.0421	0.1641	0.4961	1.0000

Focussing on the correlations between SO2 and the six other variables, we see that the correlation for SO2 and precip is very small and that for SO2 and predays is moderate. But relevant panels in the scatterplot indicate that the correlation coefficient that assesses only the linear relationship between

```
R> pairs(USairpollution,
+        panel = function (x, y, ...) {
+            points(x, y, ...)
+            abline(lm(y ~ x), col = "grey")
+        }, pch = ".", cex = 1.5)
```

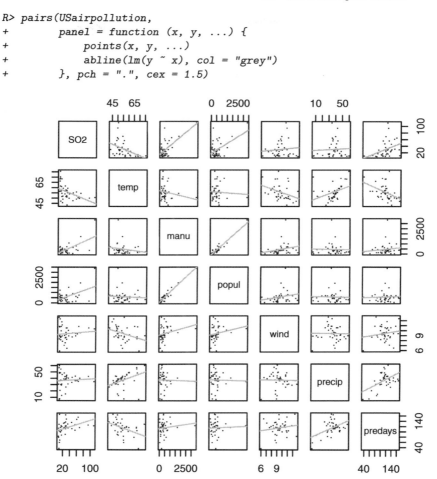

Fig. 2.11. Scatterplot matrix of the air pollution data showing the linear fit of each pair of variables.

two variables may not be suitable here and that in a multiple linear regression model for the data quadratic effects of predays and precip might be considered.

2.5 Enhancing the scatterplot with estimated bivariate densities

As we have seen above, scatterplots and scatterplot matrices are good at highlighting outliers in a multivariate data set. But in many situations another aim in examining scatterplots is to identify regions in the plot where there are high or low densities of observations that may indicate the presence of distinct groups of observations; i.e., "clusters" (see Chapter 6). But humans are not particularly good at visually examining point density, and it is often a very helpful aid to add some type of *bivariate density estimate* to the scatterplot. A bivariate density estimate is simply an approximation to the bivariate probability density function of two variables obtained from a sample of bivariate observations of the variables. If, of course, we are willing to assume a particular form of the bivariate density of the two variables, for example the bivariate normal, then estimating the density is reduced to estimating the parameters of the assumed distribution. More commonly, however, we wish to allow the data to speak for themselves and so we need to look for a *non-parametric estimation* procedure. The simplest such estimator would be a two-dimensional histogram, but for small and moderately sized data sets that is not of any real use for estimating the bivariate density function simply because most of the "boxes" in the histogram will contain too few observations; and if the number of boxes is reduced, the resulting histogram will be too coarse a representation of the density function.

Other non-parametric density estimators attempt to overcome the deficiencies of the simple two-dimensional histogram estimates by "smoothing" them in one way or another. A variety of non-parametric estimation procedures have been suggested, and they are described in detail in Silverman (1986) and Wand and Jones (1995). Here we give a brief description of just one popular class of estimators, namely *kernel density estimators*.

2.5.1 Kernel density estimators

From the definition of a probability density, if the random variable X has a density f,

$$f(x) = \lim_{h \to 0} \frac{1}{2h} P(x - h < X < x + h). \tag{2.1}$$

For any given h, a naïve estimator of $P(x - h < X < x + h)$ is the proportion of the observations $x_1, x_2, \ldots, x_n$ falling in the interval $(x - h, x + h)$,

$$\hat{f}(x) = \frac{1}{2hn} \sum_{i=1}^{n} I(x_i \in (x - h, x + h)); \tag{2.2}$$

i.e., the number of $x_1, \ldots, x_n$ falling in the interval $(x - h, x + h)$ divided by $2hn$. If we introduce a weight function W given by

$$W(x) = \begin{cases} \frac{1}{2} & |x| < 1 \\ 0 & \text{else,} \end{cases}$$

then the naïve estimator can be rewritten as

$$\hat{f}(x) = \frac{1}{n} \sum_{i=1}^{n} \frac{1}{h} W\left(\frac{x - x_i}{h}\right). \tag{2.3}$$

Unfortunately, this estimator is not a continuous function and is not particularly satisfactory for practical density estimation. It does, however, lead naturally to the kernel estimator defined by

$$\hat{f}(x) = \frac{1}{hn} \sum_{i=1}^{n} K\left(\frac{x - x_i}{h}\right), \tag{2.4}$$

where K is known as the *kernel function* and h is the *bandwidth* or *smoothing parameter*. The kernel function must satisfy the condition

$$\int_{-\infty}^{\infty} K(x)dx = 1.$$

Usually, but not always, the kernel function will be a symmetric density function; for example, the normal. Three commonly used kernel functions are rectangular,

$$K(x) = \begin{cases} \frac{1}{2} & |x| < 1 \\ 0 & \text{else.} \end{cases}$$

triangular,

$$K(x) = \begin{cases} 1 - |x| & |x| < 1 \\ 0 & \text{else,} \end{cases}$$

Gaussian,

$$K(x) = \frac{1}{\sqrt{2\pi}} e^{-\frac{1}{2}x^2}.$$

The three kernel functions are implemented in R as shown in Figure 2.12. For some grid x, the kernel functions are plotted using the R statements in Figure 2.12.

The kernel estimator $\hat{f}$ is a sum of "bumps" placed at the observations. The kernel function determines the shape of the bumps, while the window width h determines their width. Figure 2.13 (redrawn from a similar plot in Silverman 1986) shows the individual bumps $n^{-1}h^{-1}K((x - x_i)/h)$ as well as the estimate $\hat{f}$ obtained by adding them up for an artificial set of data points

```
R> rec <- function(x) (abs(x) < 1) * 0.5
R> tri <- function(x) (abs(x) < 1) * (1 - abs(x))
R> gauss <- function(x) 1/sqrt(2*pi) * exp(-(x^2)/2)
R> x <- seq(from = -3, to = 3, by = 0.001)
R> plot(x, rec(x), type = "l", ylim = c(0,1), lty = 1,
+        ylab = expression(K(x)))
R> lines(x, tri(x), lty = 2)
R> lines(x, gauss(x), lty = 3)
R> legend("topleft", legend = c("Rectangular", "Triangular",
+          "Gaussian"), lty = 1:3, title = "kernel functions",
+          bty = "n")
```

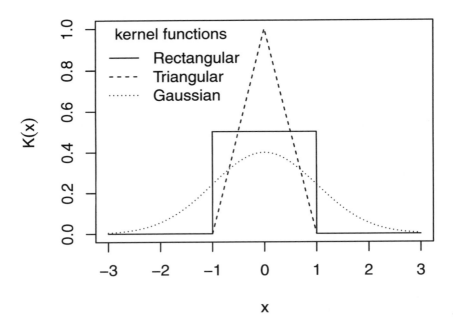

Fig. 2.12. Three commonly used kernel functions.

```
R> x <- c(0, 1, 1.1, 1.5, 1.9, 2.8, 2.9, 3.5)
R> n <- length(x)
```

For a grid

```
R> xgrid <- seq(from = min(x) - 1, to = max(x) + 1, by = 0.01)
```

on the real line, we can compute the contribution of each measurement in x, with $h = 0.4$, by the Gaussian kernel (defined in Figure 2.12, line 3) as follows:

```
R> h <- 0.4
R> bumps <- sapply(x, function(a) gauss((xgrid - a)/h)/(n * h))
```

A plot of the individual bumps and their sum, the kernel density estimate $\hat{f}$, is shown in Figure 2.13.

```
R> plot(xgrid, rowSums(bumps), ylab = expression(hat(f)(x)),
+        type = "l", xlab = "x", lwd = 2)
R> rug(x, lwd = 2)
R> out <- apply(bumps, 2, function(b) lines(xgrid, b))
```

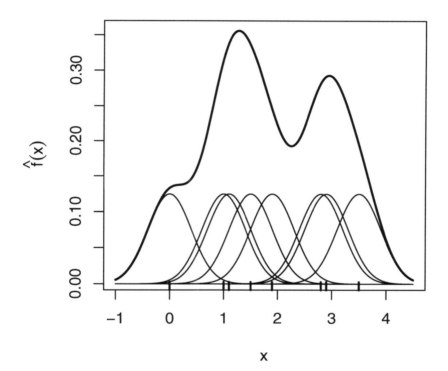

Fig. 2.13. Kernel estimate showing the contributions of Gaussian kernels evaluated for the individual observations with bandwidth $h = 0.4$.

The kernel density estimator considered as a sum of "bumps" centred at the observations has a simple extension to two dimensions (and similarly for more than two dimensions). The bivariate estimator for data (x_1, y_1), (x_2, y_2), ..., (x_n, y_n) is defined as

$$\hat{f}(x,y) = \frac{1}{nh_xh_y} \sum_{i=1}^{n} K\left(\frac{x - x_i}{h_x}, \frac{y - y_i}{h_y}\right).$$ (2.5)

In this estimator, each coordinate direction has its own smoothing parameter, h_x or h_y. An alternative is to scale the data equally for both dimensions and use a single smoothing parameter.

For bivariate density estimation, a commonly used kernel function is the standard bivariate normal density

$$K(x,y) = \frac{1}{2\pi}e^{-\frac{1}{2}(x^2+y^2)}.$$

Another possibility is the bivariate Epanechnikov kernel given by

$$K(x,y) = \begin{cases} \frac{2}{\pi}(1 - x^2 - y^2) & x^2 + y^2 < 1 \\ 0 & \text{else}, \end{cases}$$

which is implemented and depicted in Figure 2.14 by using the **persp** function for plotting in three dimensions.

According to Venables and Ripley (2002), the bandwidth should be chosen to be proportional to $n^{-1/5}$; unfortunately, the constant of proportionality depends on the unknown density. The tricky problem of bandwidth estimation is considered in detail in Silverman (1986).

Our first illustration of enhancing a scatterplot with an estimated bivariate density will involve data from the Hertzsprung-Russell (H-R) diagram of the star cluster CYG OB1, calibrated according to Vanisma and De Greve (1972). The H-R diagram is the basis of the theory of stellar evolution and is essentially a plot of the energy output of stars as measured by the logarithm of their light intensity plotted against the logarithm of their surface temperature. Part of the data is shown in Table 2.1. A scatterplot of the data enhanced by the contours of the estimated bivariate density (Wand and Ripley 2010, obtained with the function bkde2D() from the package **KernSmooth**) is shown in Figure 2.15. The plot shows the presence of two distinct clusters of stars: the larger cluster consists of stars that have high surface temperatures and a range of light intensities, and the smaller cluster contains stars with low surface temperatures and high light intensities. The bivariate density estimate can also be displayed by means of a perspective plot rather than a contour plot, and this is shown in Figure 2.16. This again demonstrates that there are two groups of stars.

Table 2.1: CYGOB1 data. Energy output and surface temperature of star cluster CYG OB1.

logst	logli	logst	logli	logst	logli
4.37	5.23	4.23	3.94	4.45	5.22

Table 2.1: CYGOB1 data (continued).

logst	logli	logst	logli	logst	logli
4.56	5.74	4.42	4.18	3.49	6.29
4.26	4.93	4.23	4.18	4.23	4.34
4.56	5.74	3.49	5.89	4.62	5.62
4.30	5.19	4.29	4.38	4.53	5.10
4.46	5.46	4.29	4.22	4.45	5.22
3.84	4.65	4.42	4.42	4.53	5.18
4.57	5.27	4.49	4.85	4.43	5.57
4.26	5.57	4.38	5.02	4.38	4.62
4.37	5.12	4.42	4.66	4.45	5.06
3.49	5.73	4.29	4.66	4.50	5.34
4.43	5.45	4.38	4.90	4.45	5.34
4.48	5.42	4.22	4.39	4.55	5.54
4.01	4.05	3.48	6.05	4.45	4.98
4.29	4.26	4.38	4.42	4.42	4.50
4.42	4.58	4.56	5.10		

For our next example of adding estimated bivariate densities to scatter-plots, we will use the body measurement data introduced in Chapter 1 (see Table 1.2), although there are rather too few observations on which to base the estimation. (The gender of each individual will not be used.) And in this case we will add the appropriate density estimate to each panel of the scatter-plot matrix of the chest, waist, and hips measurements. The resulting plot is shown in Figure 2.17. The waist/hips panel gives some evidence that there might be two groups in the data, which, of course, we know to be true, the groups being men and women. And the Waist histogram on the diagonal panel is also *bimodal*, underlining the two-group nature of the data.

2.6 Three-dimensional plots

The scatterplot matrix allows us to display information about the univariate distributions of each variable (using histograms on the main diagonal, for example) and about the bivariate distribution of all pairs of variables in a set of multivariate data. But we should perhaps consider whether the use of three-dimensional plots offers any advantage over the series of two-dimensional scatterplots used in a scatterplot matrix. To begin, we can take a look at the three-dimensional plot of the body measurements data; a version of the plot that includes simply the points along with vertical lines dropped from each point to the *x-y* plane is shown in Figure 2.18. The plot, produced with the **scatterplot3d** package (Ligges 2010), suggests the presence of two relatively separate groups of points corresponding to the males and females in the data.

```
R> epa <- function(x, y)
+       ((x^2 + y^2) < 1) * 2/pi * (1 - x^2 - y^2)
R> x <- seq(from = -1.1, to = 1.1, by = 0.05)
R> epavals <- sapply(x, function(a) epa(a, x))
R> persp(x = x, y = x, z = epavals, xlab = "x", ylab = "y",
+        zlab = expression(K(x, y)), theta = -35, axes = TRUE,
+        box = TRUE)
```

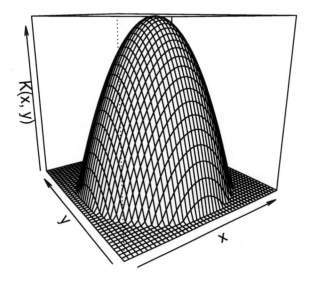

Fig. 2.14. Epanechnikov kernel for a grid between $(-1.1, -1.1)$ and $(1.1, 1.1)$.

As a second example of using a three-dimensional plot, we can look at temp, wind, and SO2 from the air pollution data. The points and vertical lines versions of the required three-dimensional plot are shown in Figure 2.19. Two observations with high SO2 levels stand out, but the plot does not appear to add much to the bubble plot for the same three variables (Figure 2.7).

Three-dimensional plots based on the original variables can be useful in some cases but may not add very much to, say, the bubble plot of the scatterplot matrix of the data. When there are many variables in a multivariate

```
R> library("KernSmooth")
R> CYGOB1d <- bkde2D(CYGOB1, bandwidth = sapply(CYGOB1, dpik))
R> plot(CYGOB1, xlab = "log surface temperature",
+                ylab = "log light intensity")
R> contour(x = CYGOB1d$x1, y = CYGOB1d$x2,
+              z = CYGOB1d$fhat, add = TRUE)
```

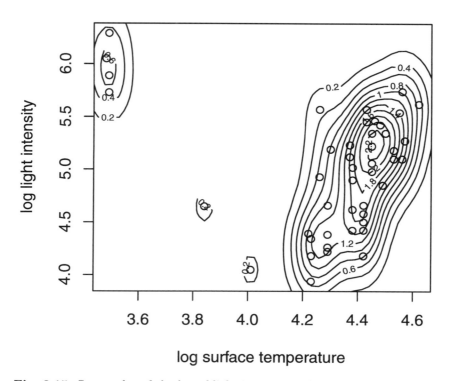

Fig. 2.15. Scatterplot of the log of light intensity and log of surface temperature for the stars in star cluster CYG OB1 showing the estimated bivariate density.

data set, there will be many possible three-dimensional plots to look at and integrating and linking all the plots may be very difficult. But if the dimensionality of the data could be reduced in some way with little loss of information, three-dimensional plots might become more useful, a point to which we will return in the next chapter.

```
R> persp(x = CYGOB1d$x1, y = CYGOB1d$x2, z = CYGOB1d$fhat,
+        xlab = "log surface temperature",
+        ylab = "log light intensity",
+        zlab = "density")
```

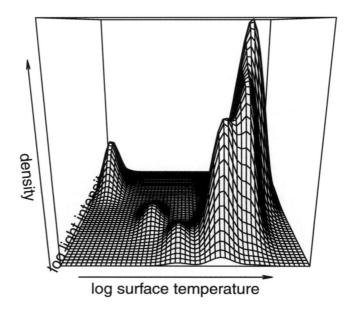

Fig. 2.16. Perspective plot of estimated bivariate density.

2.7 Trellis graphics

Trellis graphics (see Becker, Cleveland, Shyu, and Kaluzny 1994) is an approach to examining high-dimensional structure in data by means of one-, two-, and three-dimensional graphs. The problem addressed is how observations of one or more variables depend on the observations of the other variables. The essential feature of this approach is the *multiple conditioning* that allows some type of plot to be displayed for different values of a given variable (or variables). The aim is to help in understanding both the structure of the data and how well proposed models describe the structure. An example of the application of trellis graphics is given in Verbyla, Cullis, Kenward, and

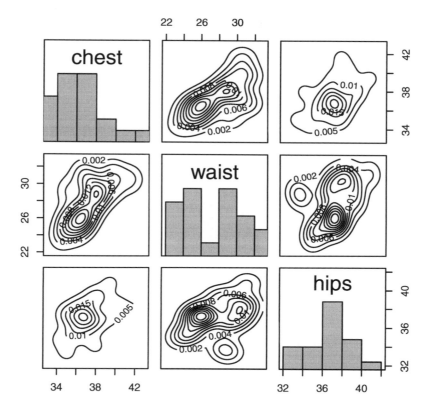

Fig. 2.17. Scatterplot matrix of body measurements data showing the estimated bivariate densities on each panel.

Welham (1999). With the recent publication of Sarkar's excellent book (see Sarkar 2008) and the development of the **lattice** (Sarkar 2010) package, trellis graphics are likely to become more popular, and in this section we will illustrate their use on multivariate data.

For the first example, we return to the air pollution data and the `temp`, `wind`, and `SO2` variables used previously to produce scatterplots of `SO2` and `temp` conditioned on values of `wind` divided into two equal parts that we shall creatively label "Light" and "High". The resulting plot is shown in Figure 2.20. The plot suggests that in cities with light winds, air pollution decreases with increasing temperature, but in cities with high winds, air pollution does not appear to be strongly related to temperature.

A more complex example of trellis graphics is shown in Figure 2.21. Here three-dimensional plots of `temp`, `wind`, and `precip` are shown for four levels of `SO2`. The graphic looks pretty, but does it convey anything of interest about

```
R> library("scatterplot3d")
R> with(measure, scatterplot3d(chest, waist, hips,
+        pch = (1:2)[gender], type = "h", angle = 55))
```

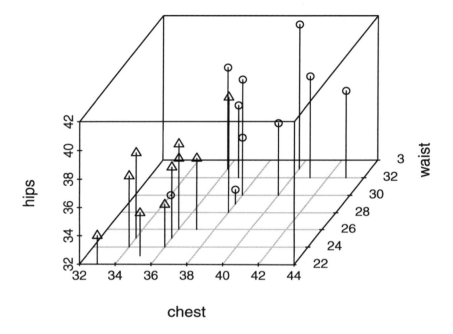

Fig. 2.18. A three-dimensional scatterplot for the body measurements data with points corresponding to male and triangles to female measurements.

the data? Probably not, as there are few points in each of the three, three-dimensional displays. This is often a problem with multipanel plots when the sample size is not large.

For the last example in this section, we will use a larger data set, namely data on earthquakes given in Sarkar (2008). The data consist of recordings of the location (latitude, longitude, and depth) and magnitude of 1000 seismic events around Fiji since 1964.

In Figure 2.22, scatterplots of latitude and longitude are plotted for three ranges of depth. The distribution of locations in the latitude-longitude space is seen to be different in the three panels, particularly for very deep quakes. In

```
R> with(USairpollution,
+       scatterplot3d(temp, wind, SO2, type = "h",
+                     angle = 55))
```

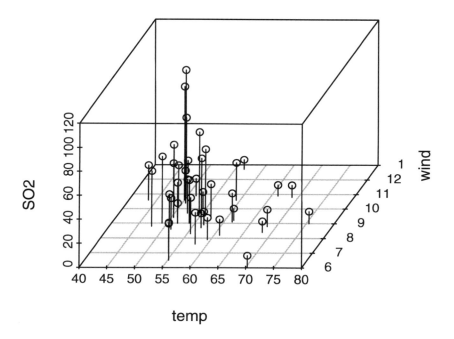

Fig. 2.19. A three-dimensional scatterplot for the air pollution data.

Figure 2.23 (a tour de force by Sarkar) the four panels are defined by ranges of magnitude and depth is encoded by different shading.

Finally, in Figure 2.24, three-dimensional scatterplots of earthquake epicentres (latitude, longitude, and depth) are plotted conditioned on earthquake magnitude. (Figures 2.22, 2.23, and 2.24 are reproduced with the kind permission of Dr. Deepayan Sarkar.)

2.8 Stalactite plots

In this section, we will describe a multivariate graphic, the stalactite plot, specifically designed for the detection and identification of multivariate out-

```
R> plot(xyplot(SO2 ~ temp| cut(wind, 2), data = USairpollution))
```

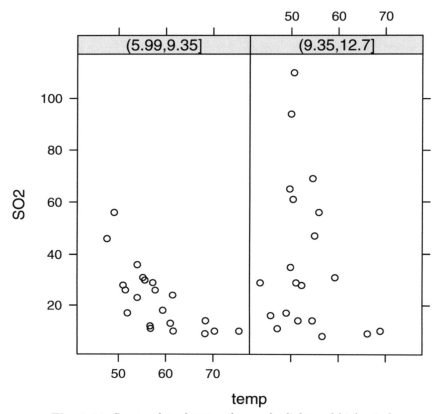

Fig. 2.20. Scatterplot of SO2 and `temp` for light and high winds.

liers. Like the chi-square plot for assessing multivariate normality, described
in Chapter 1, the stalactite plot is based on the generalised distances of observations from the multivariate mean of the data. But here these distances
are calculated from the means and covariances estimated from increasing-
sized subsets of the data. As mentioned previously when describing bivariate
boxplots, the aim is to reduce the masking effects that can arise due to the
influence of outliers on the estimates of means and covariances obtained from
all the data. The central idea of this approach is that, given distances using,
say, m observations for estimation of means and covariances, the $m + 1$ observations to be used for this estimation in the next stage are chosen to be those
with the $m + 1$ smallest distances. Thus an observation can be included in the
subset used for estimation for some value of m but can later be excluded as m
increases. Initially m is chosen to take the value $q + 1$, where q is the number
of variables in the multivariate data set because this is the smallest number

```
R> pollution <- with(USairpollution, equal.count(SO2,4))
R> plot(cloud(precip ~ temp * wind | pollution, panel.aspect = 0.9,
+        data = USairpollution))
```

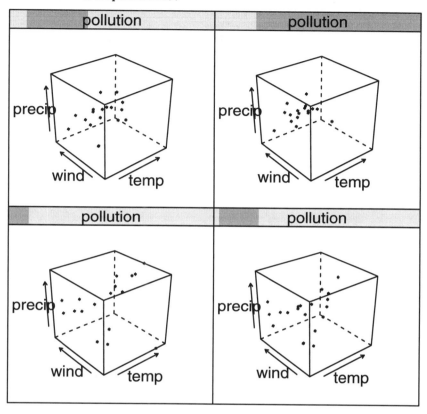

Fig. 2.21. Three-dimensional plots of `temp`, `wind`, and `precip` conditioned on levels of SO2.

allowing the calculation of the required generalised distances. The cutoff distance generally employed to identify an outlier is the maximum expected value from a sample of n random variables each having a chi-squared distribution on q degrees of freedom. The stalactite plot graphically illustrates the evolution of the outliers as the size of the subset of observations used for estimation increases. We will now illustrate the application of the stalactite plot on the US cities air pollution data. The plot (produced via `stalac(USairpollution)`) is shown in Figure 2.25. Initially most cities are indicated as outliers (a "*" in the plot), but as the number of observations on which the generalised distances are calculated is increased, the number of outliers indicated by the plot decreases. The plot clearly shows the outlying nature of a number of cities over

```
R> plot(xyplot(lat ~ long| cut(depth, 3), data = quakes,
+              layout = c(3, 1), xlab = "Longitude",
+              ylab = "Latitude"))
```

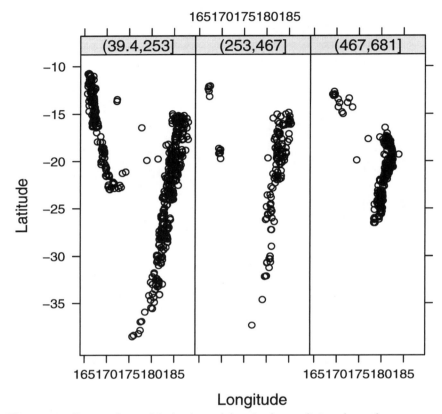

Fig. 2.22. Scatterplots of latitude and longitude conditioned on three ranges of depth.

nearly all values of m. The effect of masking is also clear; when all 41 observations are used to calculate the generalised distances, only observations Chicago, Phoenix, and Providence are indicated to be outliers.

2.9 Summary

Plotting multivariate data is an essential first step in trying to understand the story they may have to tell. The methods covered in this chapter provide just some basic ideas for taking an initial look at the data, and with software such as R there are many other possibilities for graphing multivariate obser-

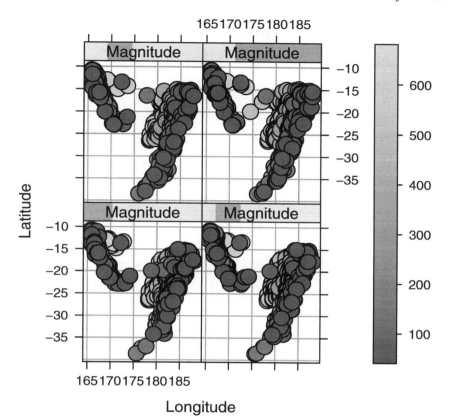

Fig. 2.23. Scatterplots of latitude and longitude conditioned on magnitude, with depth coded by shading.

vations, and readers are encouraged to explore more fully what is available. But graphics can often flatter to deceive and it is important not to be seduced when looking at a graphic into responding "what a great graph" rather than "what interesting data". A graph that calls attention to itself pictorially is almost surely a failure (see Becker et al. 1994), and unless graphs are relatively simple, they are unlikely to survive the first glance. Three-dimensional plots and trellis plots provide great pictures, which may often also be very informative (as the examples in Sarkar 2008, demonstrate), but for multivariate data with many variables, they may struggle. In many situations, the most useful graphic for a set of multivariate data may be the scatterplot matrix, perhaps with the panels enhanced in some way; for example, by the addition of bivariate density estimates or bivariate boxplots. And all the graphical approaches discussed in this chapter may become more helpful when applied to the data

```
R> plot(cloud(depth ~ lat * long | Magnitude, data = quakes,
+          zlim = rev(range(quakes$depth)),
+          screen = list(z = 105, x = -70), panel.aspect = 0.9,
+          xlab = "Longitude", ylab = "Latitude", zlab = "Depth"))
```

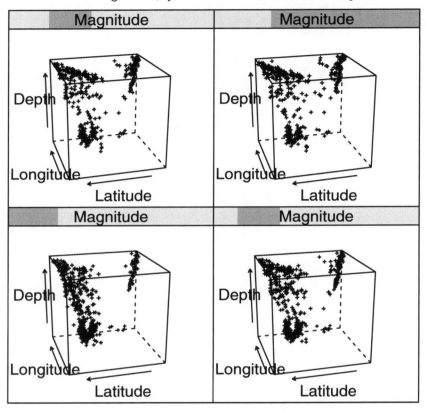

Fig. 2.24. Scatterplots of latitude and longitude conditioned on magnitude.

after their dimensionality has been reduced in some way, often by the method to be described in the next chapter.

Number of observations used for estimation

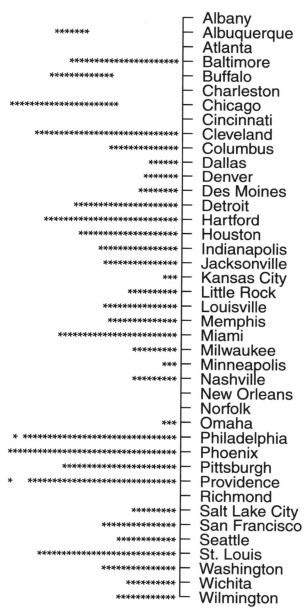

Fig. 2.25. Stalactite plot of US cities air pollution data.

2.10 Exercises

Ex. 2.1 Use the bivariate boxplot on the scatterplot of each pair of variables in the air pollution data to identify any outliers. Calculate the correlation between each pair of variables using all the data and the data with any identified outliers removed. Comment on the results.

Ex. 2.2 Compare the chi-plots with the corresponding scatterplots for each pair of variables in the air pollution data. Do you think that there is any advantage in the former?

Ex. 2.3 Construct a scatterplot matrix of the body measurements data that has the appropriate boxplot on the diagonal panels and bivariate boxplots on the other panels. Compare the plot with Figure 2.17, and say which diagram you find more informative about the data.

Ex. 2.4 Construct a further scatterplot matrix of the body measurements data that labels each point in a panel with the gender of the individual, and plot on each scatterplot the separate estimated bivariate densities for men and women.

Ex. 2.5 Construct a scatterplot matrix of the chemical composition of Romano-British pottery given in Chapter 1 (Table 1.3), identifying each unit by its kiln number and showing the estimated bivariate density on each panel. What does the resulting diagram tell you?

Ex. 2.6 Construct a bubble plot of the earthquake data using latitude and longitude as the scatterplot and depth as the circles, with greater depths giving smaller circles. In addition, divide the magnitudes into three equal ranges and label the points in your bubble plot with a different symbol depending on the magnitude group into which the point falls.

3

Principal Components Analysis

3.1 Introduction

One of the problems with a lot of sets of multivariate data is that there are simply too many variables to make the application of the graphical techniques described in the previous chapters successful in providing an informative initial assessment of the data. And having too many variables can also cause problems for other multivariate techniques that the researcher may want to apply to the data. The possible problem of too many variables is sometimes known as the *curse of dimensionality* (Bellman 1961). Clearly the scatterplots, scatterplot matrices, and other graphics included in Chapter 2 are likely to be more useful when the number of variables in the data, the *dimensionality* of the data, is relatively small rather than large. This brings us to *principal components analysis*, a multivariate technique with the central aim of reducing the dimensionality of a multivariate data set while accounting for as much of the original variation as possible present in the data set. This aim is achieved by transforming to a new set of variables, the *principal components*, that are *linear combinations* of the original variables, which are uncorrelated and are ordered so that the first *few* of them account for most of the variation in *all* the original variables. In the best of all possible worlds, the result of a principal components analysis would be the creation of a *small* number of new variables that can be used as surrogates for the originally large number of variables and consequently provide a simpler basis for, say, graphing or summarising the data, and also perhaps when undertaking further multivariate analyses of the data.

3.2 Principal components analysis (PCA)

The basic goal of principal components analysis is to describe variation in a set of correlated variables, $\mathbf{x}^\top = (x_1, \ldots, x_q)$, in terms of a new set of uncorrelated variables, $\mathbf{y}^\top = (y_1, \ldots, y_q)$, each of which is a linear combination of the **x** variables. The new variables are derived in decreasing order of "importance" in the sense that y_1 accounts for as much as possible of the variation in the original data amongst all linear combinations of **x**. Then y_2 is chosen to account for as much as possible of the remaining variation, subject to being uncorrelated with y_1, and so on. The new variables defined by this process, $y_1, \ldots, y_q$, are the principal components.

The general hope of principal components analysis is that the first few components will account for a substantial proportion of the variation in the original variables, $x_1, \ldots, x_q$, and can, consequently, be used to provide a convenient lower-dimensional summary of these variables that might prove useful for a variety of reasons. Consider, for example, a set of data consisting of examination scores for several different subjects for each of a number of students. One question of interest might be how best to construct an informative index of overall examination performance. One obvious possibility would be the mean score for each student, although if the possible or observed range of examination scores varied from subject to subject, it might be more sensible to weight the scores in some way before calculating the average, or alternatively standardise the results for the separate examinations before attempting to combine them. In this way, it might be possible to spread the students out further and so obtain a better ranking. The same result could often be achieved by applying principal components to the observed examination results and using the student's scores on the first principal components to provide a measure of examination success that maximally discriminates between them.

A further possible application for principal components analysis arises in the field of economics, where complex data are often summarised by some kind of index number; for example, indices of prices, wage rates, cost of living, and so on. When assessing changes in prices over time, the economist will wish to allow for the fact that prices of some commodities are more variable than others, or that the prices of some of the commodities are considered more important than others; in each case the index will need to be weighted accordingly. In such examples, the first principal component can often satisfy the investigator's requirements.

But it is not always the first principal component that is of most interest to a researcher. A taxonomist, for example, when investigating variation in morphological measurements on animals for which all the pairwise correlations are likely to be positive, will often be more concerned with the second and subsequent components since these might provide a convenient description of aspects of an animal's "shape". The latter will often be of more interest to the researcher than aspects of an animal's "size" which here, because of the positive correlations, will be reflected in the first principal component. For essentially the same reasons, the first principal component derived from, say, clinical psychiatric scores on patients may only provide an index of the severity of symptoms, and it is the remaining components that will give the psychiatrist important information about the "pattern" of symptoms.

The principal components are most commonly (and properly) used as a means of constructing an informative graphical representation of the data (see later in the chapter) or as input to some other analysis. One example of the latter is provided by regression analysis; principal components may be useful here when:

- There are too many explanatory variables relative to the number of observations.
- The explanatory variables are highly correlated.

Both situations lead to problems when applying regression techniques, problems that may be overcome by replacing the original explanatory variables with the first few principal component variables derived from them. An example will be given later, and other applications of the technique are described in Rencher (2002).

In some disciplines, particularly psychology and other behavioural sciences, the principal components may be considered an end in themselves and researchers may then try to interpret them in a similar fashion as for the factors in an *exploratory factor analysis* (see Chapter 5). We shall make some comments about this practise later in the chapter.

3.3 Finding the sample principal components

Principal components analysis is overwhelmingly an *exploratory* technique for multivariate data. Although there are inferential methods for using the sample principal components derived from a random sample of individuals from some population to test hypotheses about population principal components (see Jolliffe 2002), they are very rarely seen in accounts of principal components analysis that appear in the literature. Quintessentially principal components analysis is an aid for helping to understand the observed data set whether or not this is actually a "sample" in any real sense. We use this observation as the rationale for describing only *sample* principal components in this chapter.

The first principal component of the observations is that linear combination of the original variables whose sample variance is greatest amongst all possible such linear combinations. The second principal component is defined as that linear combination of the original variables that accounts for a maximal proportion of the remaining variance subject to being uncorrelated with the first principal component. Subsequent components are defined similarly. The question now arises as to how the coefficients specifying the linear combinations of the original variables defining each component are found. A little technical material is needed to answer this question.

The first principal component of the observations, y_1, is the linear combination

$$y_1 = a_{11}x_1 + a_{12}x_2 + \cdots + a_{1q}x_q$$

whose sample variance is greatest among all such linear combinations. Because the variance of y_1 could be increased without limit simply by increasing the coefficients $\mathbf{a}_1^\top = (a_{11}, a_{12}, \ldots, a_{1q})$, a restriction must be placed on these coefficients. As we shall see later, a sensible constraint is to require that the sum of squares of the coefficients should take the value one, although other constraints are possible and any multiple of the vector $\mathbf{a}_1$ produces basically the same component. To find the coefficients defining the first principal component, we need to choose the elements of the vector $\mathbf{a}_1$ so as to maximise the variance of y_1 subject to the sum of squares constraint, which can be written $\mathbf{a}_1^\top \mathbf{a}_1 = 1$. The sample variance of y_1 that is a linear function of the x variables is given by (see Chapter 1) $\mathbf{a}_1^\top \mathbf{S} \mathbf{a}_1$, where $\mathbf{S}$ is the $q \times q$ sample covariance matrix of the x variables. To maximise a function of several variables subject to one or more constraints, the method of *Lagrange multipliers* is used. Full details are given in Morrison (1990) and Jolliffe (2002), and we will not give them here. (The algebra of an example with $q = 2$ is, however, given in Section 3.5.) We simply state that the Lagrange multiplier approach leads to the solution that $\mathbf{a}_1$ is the *eigenvector* or *characteristic vector* of the sample covariance matrix, $\mathbf{S}$, corresponding to this matrix's largest *eigenvalue* or *characteristic root*. The eigenvalues λ and eigenvectors $\boldsymbol{\gamma}$ of a $q \times q$ matrix $\mathbf{A}$ are such that $\mathbf{A}\boldsymbol{\gamma} = \lambda\boldsymbol{\gamma}$; for more details, see, for example, Mardia, Kent, and Bibby (1979).

The second principal component, y_2, is defined to be the linear combination

$$y_2 = a_{21}x_1 + a_{22}x_2 + \cdots + a_{2q}x_q$$

(i.e., $y_2 = \mathbf{a}_2^\top \mathbf{x}$, where $\mathbf{a}_2^\top = (a_{21}, a_{22}, \ldots, a_{2q})$ and $\mathbf{x}^\top = (x_1, x_2, \ldots, x_q)$) that has the greatest variance subject to the following two conditions:

$$\mathbf{a}_2^\top \mathbf{a}_2 = 1,$$
$$\mathbf{a}_2^\top \mathbf{a}_1 = 0.$$

(The second condition ensures that y_1 and y_2 are uncorrelated; i.e., that the sample correlation is zero.)

Similarly, the jth principal component is that linear combination $y_j = \mathbf{a}_j^\top \mathbf{x}$ that has the greatest sample variance subject to the conditions

$$\mathbf{a}_j^\top \mathbf{a}_j = 1,$$
$$\mathbf{a}_j^\top \mathbf{a}_i = 0 \ (i < j).$$

Application of the Lagrange multiplier technique demonstrates that the vector of coefficients defining the jth principal component, $\mathbf{a}_j$, is the eigenvector of $\mathbf{S}$ associated with its jth largest eigenvalue. If the q eigenvalues of $\mathbf{S}$ are denoted by $\lambda_1, \lambda_2, \ldots, \lambda_q$, then by requiring that $\mathbf{a}_i^\top \mathbf{a}_i = 1$ it can be shown that the variance of the ith principal component is given by λ_i. The total variance of the q principal components will equal the total variance of the original variables so that

$$\sum_{i=1}^{q} \lambda_i = s_1^2 + s_2^2 + \cdots + s_q^2,$$

where s_i^2 is the sample variance of x_i. We can write this more concisely as $\sum_{i=1}^{q} \lambda_i = \text{trace}(\mathbf{S})$.

Consequently, the jth principal component accounts for a proportion P_j of the total variation of the original data, where

$$P_j = \frac{\lambda_j}{\text{trace}(\mathbf{S})}.$$

The first m principal components, where $m < q$ account for a proportion $P^{(m)}$ of the total variation in the original data, where

$$P^{(m)} = \frac{\sum_{j=1}^{m} \lambda_j}{\text{trace}(\mathbf{S})}.$$

In geometrical terms, it is easy to show that the first principal component defines the line of best fit (in the sense of minimising residuals orthogonal to the line) to the q-dimensional observations in the sample. These observations may therefore be represented in one dimension by taking their projection onto this line; that is, finding their first principal component score. If the observations happen to be collinear in q dimensions, this representation would account completely for the variation in the data and the sample covariance matrix would have only one non-zero eigenvalue. In practise, of course, such collinearity is extremely unlikely, and an improved representation would be given by projecting the q-dimensional observations onto the space of the best fit, this being defined by the first two principal components. Similarly, the first m components give the best fit in m dimensions. If the observations fit exactly into a space of m dimensions, it would be indicated by the presence of $q - m$ zero eigenvalues of the covariance matrix. This would imply the presence of $q - m$ linear relationships between the variables. Such constraints are sometimes referred to as *structural relationships*. In practise, in the vast majority of applications of principal components analysis, *all* the eigenvalues of the covariance matrix will be non-zero.

3.4 Should principal components be extracted from the covariance or the correlation matrix?

One problem with principal components analysis is that it is not scale-invariant. What this means can be explained using an example given in Mardia et al. (1979). Suppose the three variables in a multivariate data set are weight in pounds, height in feet, and age in years, but for some reason we would like our principal components expressed in ounces, inches, and decades. Intuitively two approaches seem feasible;

1. Multiply the variables by 16, 12, and 1/10, respectively and then carry out a principal components analysis on the covariance matrix of the three variables.
2. Carry out a principal components analysis on the covariance matrix of the original variables and then multiply the elements of the relevant component by 16, 12, and 1/10.

Unfortunately, these two procedures do not generally lead to the same result. So if we imagine a set of multivariate data where the variables are of completely different types, for example length, temperature, blood pressure, or anxiety rating, then the structure of the principal components derived from the covariance matrix will depend upon the essentially arbitrary choice of units of measurement; for example, changing the length from centimetres to inches will alter the derived components. Additionally, if there are large differences between the variances of the original variables, then those whose variances are largest will tend to dominate the early components. Principal components should only be extracted from the sample covariance matrix when all the original variables have roughly the same scale. But this is rare in practise and consequently, in practise, principal components are extracted from the correlation matrix of the variables, **R**. Extracting the components as the eigenvectors of **R** is equivalent to calculating the principal components from the original variables after each has been standardised to have unit variance. It should be noted, however, that there is rarely any simple correspondence between the components derived from **S** and those derived from **R**. And choosing to work with **R** rather than with **S** involves a definite but possibly arbitrary decision to make variables "equally important".

To demonstrate how the principal components of the covariance matrix of a data set can differ from the components extracted from the data's correlation matrix, we will use the example given in Jolliffe (2002). The data in this example consist of eight blood chemistry variables measured on 72 patients in a clinical trial. The correlation matrix of the data, together with the standard deviations of each of the eight variables, is

```
R> blood_corr

        [,1]    [,2]    [,3]    [,4]    [,5]    [,6]    [,7]    [,8]
[1,]   1.000   0.290   0.202  -0.055  -0.105  -0.252  -0.229   0.058
[2,]   0.290   1.000   0.415   0.285  -0.376  -0.349  -0.164  -0.129
[3,]   0.202   0.415   1.000   0.419  -0.521  -0.441  -0.145  -0.076
[4,]  -0.055   0.285   0.419   1.000  -0.877  -0.076   0.023  -0.131
[5,]  -0.105  -0.376  -0.521  -0.877   1.000   0.206   0.034   0.151
[6,]  -0.252  -0.349  -0.441  -0.076   0.206   1.000   0.192   0.077
[7,]  -0.229  -0.164  -0.145   0.023   0.034   0.192   1.000   0.423
[8,]   0.058  -0.129  -0.076  -0.131   0.151   0.077   0.423   1.000
```

```
R> blood_sd
```

```
rblood  plate  wblood    neut   lymph  bilir  sodium  potass
 0.371 41.253   1.935   0.077   0.071  4.037   2.732   0.297
```

There are considerable differences between these standard deviations. We can apply principal components analysis to both the covariance and correlation matrix of the data using the following R code:

```
R> blood_pcacov <- princomp(covmat = blood_cov)
R> summary(blood_pcacov, loadings = TRUE)
```

```
Importance of components:
                          Comp.1    Comp.2    Comp.3    Comp.4
Standard deviation      41.2877  3.880213  2.641973  1.624584
Proportion of Variance   0.9856  0.008705  0.004036  0.001526
Cumulative Proportion    0.9856  0.994323  0.998359  0.999885
                          Comp.5    Comp.6    Comp.7    Comp.8
Standard deviation     3.540e-01 2.562e-01 8.511e-02 2.373e-02
Proportion of Variance 7.244e-05 3.794e-05 4.188e-06 3.255e-07
Cumulative Proportion  1.000e+00 1.000e+00 1.000e+00 1.000e+00
```

```
Loadings:
       Comp.1 Comp.2 Comp.3 Comp.4 Comp.5 Comp.6 Comp.7 Comp.8
rblood                              0.943  0.329
plate  -0.999
wblood        -0.192        -0.981
neut                                              0.758 -0.650
lymph                                            -0.649 -0.760
bilir          0.961  0.195 -0.191
sodium         0.193 -0.979
potass                              0.329 -0.942
```

```
R> blood_pcacor <- princomp(covmat = blood_corr)
R> summary(blood_pcacor, loadings = TRUE)
```

```
Importance of components:
                        Comp.1 Comp.2 Comp.3 Comp.4 Comp.5
Standard deviation       1.671 1.2376 1.1177 0.8823 0.7884
Proportion of Variance   0.349 0.1915 0.1562 0.0973 0.0777
Cumulative Proportion    0.349 0.5405 0.6966 0.7939 0.8716
                        Comp.6 Comp.7  Comp.8
Standard deviation      0.6992 0.66002 0.31996
Proportion of Variance  0.0611 0.05445 0.01280
Cumulative Proportion   0.9327 0.98720 1.00000
```

```
Loadings:
```

	Comp.1	Comp.2	Comp.3	Comp.4	Comp.5	Comp.6	Comp.7	Comp.8
[1,]	-0.194	0.417	0.400	0.652	0.175	-0.363	0.176	0.102
[2,]	-0.400	0.154	0.168		-0.848	0.230	-0.110	
[3,]	-0.459		0.168	-0.274	0.251	0.403	0.677	
[4,]	-0.430	-0.472	-0.171	0.169	0.118		-0.237	0.678
[5,]	0.494	0.360		-0.180	-0.139	0.136	0.157	0.724
[6,]	0.319	-0.320	-0.277	0.633	-0.162	0.384	0.377	
[7,]	0.177	-0.535	0.410	-0.163	-0.299	-0.513	0.367	
[8,]	0.171	-0.245	0.709		0.198	0.469	-0.376	

(The "blanks" in this output represent very small values.) Examining the results, we see that each of the principal components of the covariance matrix is largely dominated by a single variable, whereas those for the correlation matrix have moderate-sized coefficients on several of the variables. And the first component from the covariance matrix accounts for almost 99% of the total variance of the observed variables. The components of the covariance matrix are completely dominated by the fact that the variance of the plate variable is roughly 400 times larger than the variance of any of the seven other variables. Consequently, the principal components from the covariance matrix simply reflect the order of the sizes of the variances of the observed variables. The results from the correlation matrix tell us, in particular, that a weighted contrast of the first four and last four variables is the linear function with the largest variance. This example illustrates that when variables are on very different scales or have very different variances, a principal components analysis of the data should be performed on the correlation matrix, *not* on the covariance matrix.

3.5 Principal components of bivariate data with correlation coefficient r

Before we move on to look at some practical examples of the application of principal components analysis, it will be helpful to look in a little more detail at the mathematics of the method in one very simple case. We will do this in this section for bivariate data where the two variables, x_1 and x_2, have correlation coefficient r. The sample correlation matrix in this case is simply

$$\mathbf{R} = \begin{pmatrix} 1.0 & r \\ r & 0.1 \end{pmatrix}.$$

In order to find the principal components of the data we need to find the eigenvalues and eigenvectors of $\mathbf{R}$. The eigenvalues are found as the roots of the equation

$$\det(\mathbf{R} - \lambda \mathbf{I}) = 0.$$

This leads to the quadratic equation in λ

$$(1 - \lambda)^2 - r^2 = 0,$$

and solving this equation leads to eigenvalues $\lambda_1 = 1 + r$, $\lambda_2 = 1 - r$. Note that the sum of the eigenvalues is two, equal to trace($\mathbf{R}$). The eigenvector corresponding to λ_1 is obtained by solving the equation

$$\mathbf{R}\mathbf{a}_1 = \lambda_1 \mathbf{a}_1.$$

This leads to the equations

$$a_{11} + ra_{12} = (1 + r)a_{11},$$
$$ra_{11} + a_{12} = (1 + r)a_{12}.$$

The two equations are identical, and both reduce to requiring $a_{11} = a_{12}$. If we now introduce the normalisation constraint $\mathbf{a}_1^\top \mathbf{a}_1 = 1$, we find that

$$a_{11} = a_{12} = \frac{1}{\sqrt{2}}.$$

Similarly, we find the second eigenvector is given by $a_{21} = \frac{1}{\sqrt{2}}$ and $a_{22} = -\frac{1}{\sqrt{2}}$. The two principal components are then given by

$$y_1 = \frac{1}{\sqrt{2}}(x_1 + x_2), \qquad y_2 = \frac{1}{\sqrt{2}}(x_1 - x_2).$$

We can calculate the sample variance of the first principal component as

$$\begin{aligned}
\mathsf{Var}(y_1) &= \mathsf{Var}\left[\frac{1}{\sqrt{2}}(x_1 + x_2)\right] = \frac{1}{2}\mathsf{Var}(x_1 + x_2) \\
&= \frac{1}{2}[\mathsf{Var}(x_1) + \mathsf{Var}(x_2) + 2\mathsf{Cov}(x_1, x_2)] \\
&= \frac{1}{2}(1 + 1 + 2r) = 1 + r.
\end{aligned}$$

Similarly, the variance of the second principal component is $1 - r$.

Notice that if $r < 0$, the order of the eigenvalues and hence of the principal components is reversed; if $r = 0$, the eigenvalues are both equal to 1 and any two solutions at right angles could be chosen to represent the two components. Two further points should be noted:

1. There is an arbitrary sign in the choice of the elements of $\mathbf{a}_i$. It is customary (but not universal) to choose a_{i1} to be positive.
2. The coefficients that define the two components do not depend on r, although the proportion of variance explained by each does change with r. As r tends to 1, the proportion of variance accounted for by y_1, namely $(1 + r)/2$, also tends to one. When $r = 1$, the points all align on a straight line and the variation in the data is unidimensional.

3.6 Rescaling the principal components

The coefficients defining the principal components derived as described in the previous section are often rescaled so that they are correlations or covariances between the original variables and the derived components. The rescaled coefficients are often useful in interpreting a principal components analysis. The covariance of variable i with component j is given by

$$\mathsf{Cov}(x_i, y_j) = \lambda_j a_{ji}.$$

The correlation of variable x_i with component y_j is therefore

$$r_{x_i, y_j} = \frac{\lambda_j a_{ji}}{\sqrt{\mathsf{Var}(x_i)\mathsf{Var}(y_j)}} = \frac{\lambda_j a_{ji}}{s_i \sqrt{\lambda_j}} = \frac{a_{ji}\sqrt{\lambda_j}}{s_i}.$$

If the components are extracted from the correlation matrix rather than the covariance matrix, the correlation between variable and component becomes

$$r_{x_i, y_j} = a_{ji}\sqrt{\lambda_j}$$

because in this case the standard deviation, s_i, is unity. (Although for convenience we have used the same nomenclature for the eigenvalues and the eigenvectors extracted from the covariance matrix or the correlation matrix, they will, of course, not be equal.) The rescaled coefficients from a principal components analysis of a correlation matrix are analogous to *factor loadings*, as we shall see in Chapter 5. Often these rescaled coefficients are presented as the results of a principal components analysis and used in interpretation.

3.7 How the principal components predict the observed covariance matrix

In this section, we will look at how the principal components reproduce the observed covariance or correlation matrix from which they were extracted. To begin, let the initial vectors $\mathbf{a}_1, \mathbf{a}_2, \ldots, \mathbf{a}_q$, that define the principal components be used to form a $q \times q$ matrix, $\mathbf{A} = (\mathbf{a}_1, \mathbf{a}_2, \ldots, \mathbf{a}_q)$; we assume that these are vectors extracted from the covariance matrix, $\mathbf{S}$, and scaled so that $\mathbf{a}_i^\top \mathbf{a}_i = 1$. Arrange the eigenvalues $\lambda_1, \lambda_2, \ldots, \lambda_q$ along the main diagonal of a diagonal matrix, $\boldsymbol{\Lambda}$. Then it can be shown that the covariance matrix of the observed variables $x_1, x_2, \ldots, x_q$ is given by

$$\mathbf{S} = \mathbf{A}\boldsymbol{\Lambda}\mathbf{A}^\top.$$

This is known as the *spectral decomposition* of $\mathbf{S}$. Rescaling the vectors $\mathbf{a}_1, \mathbf{a}_2, \ldots, \mathbf{a}_q$ so that the sum of squares of their elements is equal to the corresponding eigenvalue (i.e., calculating $\mathbf{a}_i^* = \lambda_i^{\frac{1}{2}} \mathbf{a}_i$) allows $\mathbf{S}$ to be written more simply as

$$S = A^* A^{*\top},$$

where $A^* = \begin{pmatrix} a_1^* \dots a_q^* \end{pmatrix}$.

If the matrix A_m^* is formed from, say, the first m components rather than from all q, then $A_m^* A_m^{*\top}$ gives the predicted value of S based on these m components. It is often useful to calculate such a predicted value based on the number of components considered adequate to describe the data to informally assess the "fit" of the principal components analysis. How this number of components might be chosen is considered in the next section.

3.8 Choosing the number of components

As described earlier, principal components analysis is seen to be a technique for transforming a set of observed variables into a new set of variables that are uncorrelated with one another. The variation in the original q variables is only *completely* accounted for by *all* q principal components. The usefulness of these transformed variables, however, stems from their property of accounting for the variance in decreasing proportions. The first component, for example, accounts for the maximum amount of variation possible for any linear combination of the original variables. But how useful is this artificial variate constructed from the observed variables? To answer this question we would first need to know the proportion of the total variance of the original variables for which it accounted. If, for example, 80% of the variation in a multivariate data set involving six variables could be accounted for by a simple weighted average of the variable values, then almost all the variation can be expressed along a single continuum rather than in six-dimensional space. The principal components analysis would have provided a highly parsimonious summary (reducing the dimensionality of the data from six to one) that might be useful in later analysis.

So the question we need to ask is how many components are needed to provide an adequate summary of a given data set. A number of informal and more formal techniques are available. Here we shall concentrate on the former; examples of the use of formal inferential methods are given in Jolliffe (2002) and Rencher (2002).

The most common of the relatively ad hoc procedures that have been suggested for deciding upon the number of components to retain are the following:

- Retain just enough components to explain some specified large percentage of the total variation of the original variables. Values between 70% and 90% are usually suggested, although smaller values might be appropriate as q or n, the sample size, increases.

- Exclude those principal components whose eigenvalues are less than the average, $\sum_{i=1}^{q} \frac{\lambda_i}{q}$. Since $\sum_{i=1}^{q} \lambda_i = \text{trace}(S)$, the average eigenvalue is also the average variance of the original variables. This method then retains those components that account for more variance than the average for the observed variables.

- When the components are extracted from the correlation matrix, trace($\mathbf{R}$) = q, and the average variance is therefore one, so applying the rule in the previous bullet point, components with eigenvalues less than one are excluded. This rule was originally suggested by Kaiser (1958), but Jolliffe (1972), on the basis of a number of simulation studies, proposed that a more appropriate procedure would be to exclude components extracted from a correlation matrix whose associated eigenvalues are less than 0.7.

- Cattell (1966) suggests examination of the plot of the λ_i against i, the so-called scree diagram. The number of components selected is the value of i corresponding to an "elbow" in the curve, i.e., a change of slope from "steep" to "shallow". In fact, Cattell was more specific than this, recommending to look for a point on the plot beyond which the scree diagram defines a more or less straight line, not necessarily horizontal. The first point on the straight line is then taken to be the last component to be retained. And it should also be remembered that Cattell suggested the scree diagram in the context of factor analysis rather than applied to principal components analysis.

- A modification of the scree digram described by Farmer (1971) is the *log-eigenvalue diagram* consisting of a plot of $\log(\lambda_i)$ against i.

Returning to the results of the principal components analysis of the blood chemistry data given in Section 3.3, we find that the first four components account for nearly 80% of the total variance, but it takes a further two components to push this figure up to 90%. A cutoff of one for the eigenvalues leads to retaining three components, and with a cutoff of 0.7 four components are kept. Figure 3.1 shows the scree diagram and log-eigenvalue diagram for the data and the R code used to construct the two diagrams. The former plot may suggest four components, although this is fairly subjective, and the latter seems to be of little help here because it appears to indicate retaining seven components, hardly much of a dimensionality reduction. The example illustrates that the proposed methods for deciding how many components to keep can (and often do) lead to different conclusions.

3.9 Calculating principal components scores

If we decide that we need, say, m principal components to adequately represent our data (using one or another of the methods described in the previous section), then we will generally wish to calculate the scores on each of these components for each individual in our sample. If, for example, we have derived the components from the covariance matrix, $\mathbf{S}$, then the m principal components scores for individual i with original $q \times 1$ vector of variable values x_i are obtained as

```
R> plot(blood_pcacor$sdev^2, xlab = "Component number",
+       ylab = "Component variance", type = "l", main = "Scree diagram")
R> plot(log(blood_pcacor$sdev^2), xlab = "Component number",
+       ylab = "log(Component variance)", type="l",
+       main = "Log(eigenvalue) diagram")
```

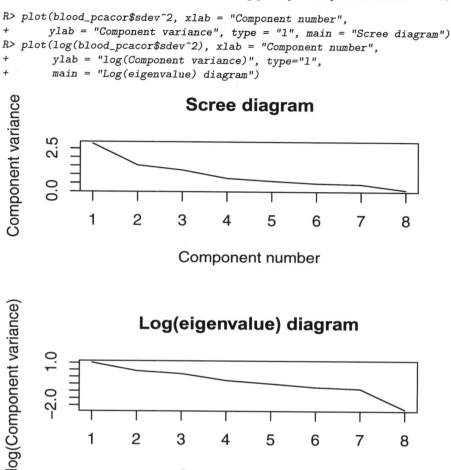

Fig. 3.1. Scree diagram and log-eigenvalue diagram for principal components of the correlation matrix of the blood chemistry data.

$$y_{i1} = \mathbf{a}_1^\top \mathbf{x}_i$$
$$y_{i2} = \mathbf{a}_2^\top \mathbf{x}_i$$
$$\vdots$$
$$y_{im} = \mathbf{a}_m^\top \mathbf{x}_i$$

If the components are derived from the correlation matrix, then $\mathbf{x}_i$ would contain individual i's standardised scores for each variable.

The principal components scores calculated as above have variances equal to λ_j for $j = 1, \ldots, m$. Many investigators might prefer to have scores with mean zero and variance equal to unity. Such scores can be found as

$$\mathbf{z} = \boldsymbol{\Lambda}_m^{-1} \mathbf{A}_m^{\top} \mathbf{x},$$

where $\boldsymbol{\Lambda}_m$ is an $m \times m$ diagonal matrix with $\lambda_1, \lambda_2, \ldots, \lambda_m$ on the main diagonal, $\mathbf{A}_m = (\mathbf{a}_1 \ldots \mathbf{a}_m)$, and $\mathbf{x}$ is the $q \times 1$ vector of standardised scores. We should note here that the first m principal components scores are the same whether we retain all possible q components or just the first m. As we shall see in Chapter 5, this is *not* the case with the calculation of factor scores.

3.10 Some examples of the application of principal components analysis

In this section, we will look at the application of PCA to a number of data sets, beginning with one involving only two variables, as this allows us to illustrate graphically an important point about this type of analysis.

3.10.1 Head lengths of first and second sons

Table 3.1: `headsize` data. Head Size Data.

head1	breadth1	head2	breadth2	head1	breadth1	head2	breadth2
191	155	179	145	190	159	195	157
195	149	201	152	188	151	187	158
181	148	185	149	163	137	161	130
183	153	188	149	195	155	183	158
176	144	171	142	186	153	173	148
208	157	192	152	181	145	182	146
189	150	190	149	175	140	165	137
197	159	189	152	192	154	185	152
188	152	197	159	174	143	178	147
192	150	187	151	176	139	176	143
179	158	186	148	197	167	200	158
183	147	174	147	190	163	187	150
174	150	185	152				

The data in Table 3.1 give the head lengths and head breadths (in millimetres) for each of the first two adult sons in 25 families. Here we shall use only the head lengths; the head breadths will be used later in the chapter. The mean vector and covariance matrix of the head length measurements are found using

```
R> head_dat <- headsize[, c("head1", "head2")]
R> colMeans(head_dat)
```

```
head1 head2
185.7 183.8
```

```
R> cov(head_dat)
```

```
        head1   head2
head1 95.29   69.66
head2 69.66 100.81
```

The principal components of these data, extracted from their covariance matrix, can be found using

```
R> head_pca <- princomp(x = head_dat)
R> head_pca
```

```
Call:
princomp(x = head_dat)
```

```
Standard deviations:
Comp.1 Comp.2
12.691  5.215
```

```
 2  variables and  25 observations.
```

```
R> print(summary(head_pca), loadings = TRUE)
```

```
Importance of components:
                        Comp.1 Comp.2
Standard deviation      12.6908 5.2154
Proportion of Variance  0.8555 0.1445
Cumulative Proportion   0.8555 1.0000
```

```
Loadings:
      Comp.1 Comp.2
head1  0.693 -0.721
head2  0.721  0.693
```

and are
$$y_1 = 0.693x_1 + 0.721x_2 \quad y_2 = -0.721x_1 + 0.693x_2$$

with variances 167.77 and 28.33. The first principal component accounts for a proportion $167.77/(167.77 + 28.33) = 0.86$ of the total variance in the original variables. Note that the total variance of the principal components is 196.10, which as expected is equal to the total variance of the original variables, found by adding the relevant terms in the covariance matrix given earlier; i.e., $95.29 + 100.81 = 196.10$.

How should the two derived components be interpreted? The first component is essentially the sum of the head lengths of the two sons, and the second component is the difference in head lengths. Perhaps we can label the first component "size" and the second component "shape", but later we will have some comments about trying to give principal components such labels.

To calculate an individual's score on a component, we simply multiply the variable values minus the appropriate mean by the loading for the variable and add these values over all variables. We can illustrate this calculation using the data for the first family, where the head length of the first son is 191 mm and for the second son 179 mm. The score for this family on the first principal component is calculated as

$$0.693 \cdot (191 - 185.72) + 0.721 \cdot (179 - 183.84) = 0.169,$$

and on the second component the score is

$$-0.721 \cdot (191 - 185.72) + 0.693 \cdot (179 - 183.84) = -7.61.$$

The variance of the first principal components scores will be 167.77, and the variance of the second principal component scores will be 28.33.

We can plot the data showing the axes corresponding to the principal components. The first axis passes through the mean of the data and has slope $0.721/0.693$, and the second axis also passes through the mean and has slope $-0.693/0.721$. The plot is shown in Figure 3.2. This example illustrates that a principal components analysis is essentially simply a rotation of the axes of the multivariate data scatter. And we can also plot the principal components scores to give Figure 3.3. (Note that in this figure the range of the x-axis and the range for the y-axis have been made the same to account for the larger variance of the first principal component.)

We can use the principal components analysis of the head size data to demonstrate how the principal components reproduce the observed covariance matrix. We first need to rescale the principal components we have at this point by multiplying them by the square roots of their respective variances to give the new components

$$y_1 = 12.952(0.693x_1 + 0.721x_2), \text{ i.e., } y_1 = 8.976x_1 + 9.338x_2$$

and

$$y_2 = 5.323(-0.721x_1 + 0.693x_2), \text{ i.e., } y_2 = -3.837x_1 + 3.688x_2,$$

leading to the matrix $\mathbf{A}_2^*$ as defined in Section 1.5.1:

$$\mathbf{A}_2^* = \begin{pmatrix} 8.976 & -3.837 \\ 9.338 & 3.688 \end{pmatrix}.$$

Multiplying this matrix by its transpose should recreate the covariance matrix of the head length data; doing the matrix multiplication shows that it does recreate $\mathbf{S}$:

```
R> a1<-183.84-0.721*185.72/0.693
R> b1<-0.721/0.693
R> a2<-183.84-(-0.693*185.72/0.721)
R> b2<--0.693/0.721
R> plot(head_dat, xlab = "First son's head length (mm)",
+         ylab = "Second son's head length")
R> abline(a1, b1)
R> abline(a2, b2, lty = 2)
```

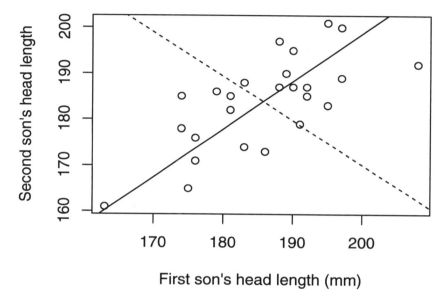

Fig. 3.2. Head length of first and second sons, showing axes corresponding to the principal components of the sample covariance matrix of the data.

$$\mathbf{A}_2^*(\mathbf{A}_2^*)^\top = \begin{pmatrix} 95.29 & 69.66 \\ 69.66 & 100.81 \end{pmatrix}.$$

(As an exercise, readers might like to find the predicted covariance matrix using only the first component.)

The head size example has been useful for discussing some aspects of principal components analysis but it is not, of course, typical of multivariate data sets encountered in practise, where many more than two variables will be recorded for each individual in a study. In the next two subsections, we consider some more interesting examples.

```
R> xlim <- range(head_pca$scores[,1])
R> plot(head_pca$scores, xlim = xlim, ylim = xlim)
```

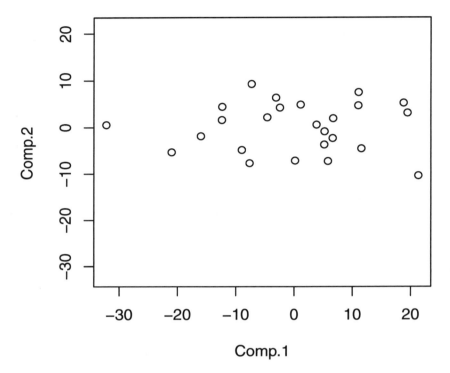

Fig. 3.3. Plot of the first two principal component scores for the head size data.

3.10.2 Olympic heptathlon results

The pentathlon for women was first held in Germany in 1928. Initially this consisted of the shot put, long jump, 100 m, high jump, and javelin events, held over two days. In the 1964 Olympic Games, the pentathlon became the first combined Olympic event for women, consisting now of the 80 m hurdles, shot, high jump, long jump, and 200 m. In 1977, the 200 m was replaced by the 800 m run, and from 1981 the IAAF brought in the seven-event heptathlon in place of the pentathlon, with day one containing the events 100 m hurdles, shot, high jump, and 200 m run, and day two the long jump, javelin, and 800 m run. A scoring system is used to assign points to the results from each event, and the winner is the woman who accumulates the most points over the two days. The event made its first Olympic appearance in 1984.

Table 3.2: heptathlon data. Results of Olympic heptathlon, Seoul, 1988.

	hurdles	highjump	shot	run200m	longjump	javelin	run800m	score
Joyner-Kersee (USA)	12.69	1.86	15.80	22.56	7.27	45.66	128.51	7291
John (GDR)	12.85	1.80	16.23	23.65	6.71	42.56	126.12	6897
Behmer (GDR)	13.20	1.83	14.20	23.10	6.68	44.54	124.20	6858
Sablovskaite (URS)	13.61	1.80	15.23	23.92	6.25	42.78	132.24	6540
Choubenkova (URS)	13.51	1.74	14.76	23.93	6.32	47.46	127.90	6540
Schulz (GDR)	13.75	1.83	13.50	24.65	6.33	42.82	125.79	6411
Fleming (AUS)	13.38	1.80	12.88	23.59	6.37	40.28	132.54	6351
Greiner (USA)	13.55	1.80	14.13	24.48	6.47	38.00	133.65	6297
Lajbnerova (CZE)	13.63	1.83	14.28	24.86	6.11	42.20	136.05	6252
Bouraga (URS)	13.25	1.77	12.62	23.59	6.28	39.06	134.74	6252
Wijnsma (HOL)	13.75	1.86	13.01	25.03	6.34	37.86	131.49	6205
Dimitrova (BUL)	13.24	1.80	12.88	23.59	6.37	40.28	132.54	6171
Scheider (SWI)	13.85	1.86	11.58	24.87	6.05	47.50	134.93	6137
Braun (FRG)	13.71	1.83	13.16	24.78	6.12	44.58	142.82	6109
Ruotsalainen (FIN)	13.79	1.80	12.32	24.61	6.08	45.44	137.06	6101
Yuping (CHN)	13.93	1.86	14.21	25.00	6.40	38.60	146.67	6087
Hagger (GB)	13.47	1.80	12.75	25.47	6.34	35.76	138.48	5975
Brown (USA)	14.07	1.83	12.69	24.83	6.13	44.34	146.43	5972
Mulliner (GB)	14.39	1.71	12.68	24.92	6.10	37.76	138.02	5746
Hautenauve (BEL)	14.04	1.77	11.81	25.61	5.99	35.68	133.90	5734
Kytola (FIN)	14.31	1.77	11.66	25.69	5.75	39.48	133.35	5686
Geremias (BRA)	14.23	1.71	12.95	25.50	5.50	39.64	144.02	5508
Hui-Ing (TAI)	14.85	1.68	10.00	25.23	5.47	39.14	137.30	5290
Jeong-Mi (KOR)	14.53	1.71	10.83	26.61	5.50	39.26	139.17	5289
Launa (PNG)	16.42	1.50	11.78	26.16	4.88	46.38	163.43	4566

In the 1988 Olympics held in Seoul, the heptathlon was won by one of the stars of women's athletics in the USA, Jackie Joyner-Kersee. The results for all 25 competitors in all seven disciplines are given in Table 3.2 (from Hand, Daly, Lunn, McConway, and Ostrowski 1994). We shall analyse these data using principal components analysis with a view to exploring the structure of the data and assessing how the derived principal components scores (discussed later) relate to the scores assigned by the official scoring system.

But before undertaking the principal components analysis, it is good data analysis practise to carry out an initial assessment of the data using one or another of the graphics described in Chapter 2. Some numerical summaries may also be helpful before we begin the main analysis. And before any of these, it will help to score all seven events in the same direction so that "large" values are indicative of a "better" performance. The R code for reversing the values for some events, then calculating the correlation coefficients between the ten events and finally constructing the scatterplot matrix of the data is

```
R> heptathlon$hurdles <- with(heptathlon, max(hurdles)-hurdles)
R> heptathlon$run200m <- with(heptathlon, max(run200m)-run200m)
R> heptathlon$run800m <- with(heptathlon, max(run800m)-run800m)
```

```
R> score <- which(colnames(heptathlon) == "score")
R> round(cor(heptathlon[,-score]), 2)
```

	hurdles	highjump	shot	run200m	longjump	javelin	run800m
hurdles	1.00	0.81	0.65	0.77	0.91	0.01	0.78
highjump	0.81	1.00	0.44	0.49	0.78	0.00	0.59
shot	0.65	0.44	1.00	0.68	0.74	0.27	0.42
run200m	0.77	0.49	0.68	1.00	0.82	0.33	0.62
longjump	0.91	0.78	0.74	0.82	1.00	0.07	0.70
javelin	0.01	0.00	0.27	0.33	0.07	1.00	-0.02
run800m	0.78	0.59	0.42	0.62	0.70	-0.02	1.00

```
R> plot(heptathlon[,-score])
```

The scatterplot matrix appears in Figure 3.4.

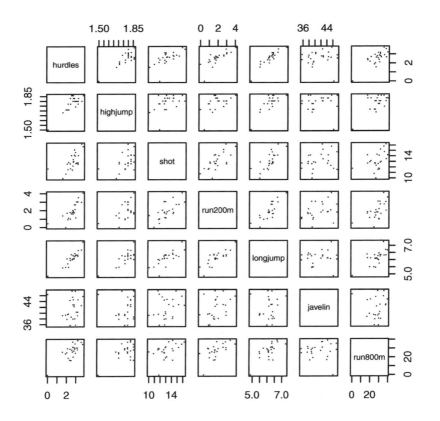

Fig. 3.4. Scatterplot matrix of the seven heptathlon events after transforming some variables so that for all events large values are indicative of a better performance.

Examination of the correlation matrix shows that most pairs of events are positively correlated, some moderately (for example, high jump and shot) and others relatively highly (for example, high jump and hurdles). The exceptions to this general observation are the relationships between the javelin event and the others, where almost all the correlations are close to zero. One explanation might be that the javelin is a very "technical" event and perhaps the training for the other events does not help the competitors in the javelin. But before we speculate further, we should look at the scatterplot matrix of the seven events shown in Figure 3.4. One very clear observation in this plot is that for all events *except* the javelin there is an outlier who is very much poorer than the other athletes at these six events, and this is the competitor from Papua New Guinea (PNG), who finished last in the competition in terms of points scored. But surprisingly, in the scatterplots involving the javelin, it is this competitor who again stands out, but in this case she has the third *highest* value for the event. It might be sensible to look again at both the correlation matrix and the scatterplot matrix after removing the competitor from PNG; the relevant R code is

```
R> heptathlon <- heptathlon[-grep("PNG", rownames(heptathlon)),]
R> score <- which(colnames(heptathlon) == "score")
R> round(cor(heptathlon[,-score]), 2)
```

	hurdles	highjump	shot	run200m	longjump	javelin	run800m
hurdles	1.00	0.58	0.77	0.83	0.89	0.33	0.56
highjump	0.58	1.00	0.46	0.39	0.66	0.35	0.15
shot	0.77	0.46	1.00	0.67	0.78	0.34	0.41
run200m	0.83	0.39	0.67	1.00	0.81	0.47	0.57
longjump	0.89	0.66	0.78	0.81	1.00	0.29	0.52
javelin	0.33	0.35	0.34	0.47	0.29	1.00	0.26
run800m	0.56	0.15	0.41	0.57	0.52	0.26	1.00

The new scatterplot matrix is shown in Figure 3.5. Several of the correlations are changed to some degree from those shown before removal of the PNG competitor, particularly the correlations involving the javelin event, where the very small correlations between performances in this event and the others have increased considerably. Given the relatively large overall change in the correlation matrix produced by omitting the PNG competitor, we shall extract the principal components of the data from the correlation matrix *after* this omission. The principal components can now be found using

```
R> heptathlon_pca <- prcomp(heptathlon[, -score], scale = TRUE)
R> print(heptathlon_pca)
```

Standard deviations:
[1] 2.08 0.95 0.91 0.68 0.55 0.34 0.26

Rotation:

```
R> plot(heptathlon[,-score], pch = ".", cex = 1.5)
```

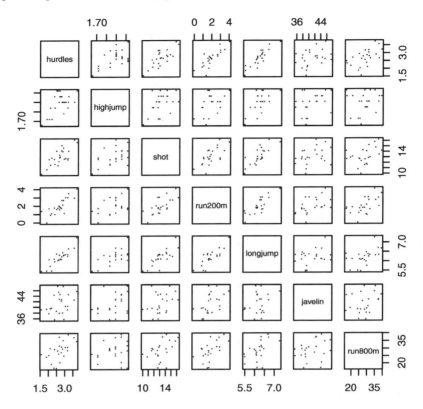

Fig. 3.5. Scatterplot matrix for the **heptathlon** data after removing observations of the PNG competitor.

	PC1	PC2	PC3	PC4	PC5	PC6	PC7
hurdles	-0.45	0.058	-0.17	0.048	-0.199	0.847	-0.070
highjump	-0.31	-0.651	-0.21	-0.557	0.071	-0.090	0.332
shot	-0.40	-0.022	-0.15	0.548	0.672	-0.099	0.229
run200m	-0.43	0.185	0.13	0.231	-0.618	-0.333	0.470
longjump	-0.45	-0.025	-0.27	-0.015	-0.122	-0.383	-0.749
javelin	-0.24	-0.326	0.88	0.060	0.079	0.072	-0.211
run800m	-0.30	0.657	0.19	-0.574	0.319	-0.052	0.077

The **summary** method can be used for further inspection of the details:

```
R> summary(heptathlon_pca)
```

```
Importance of components:
                          PC1    PC2    PC3    PC4    PC5    PC6
```

```
Standard deviation      2.079 0.948 0.911 0.6832 0.5462 0.3375
Proportion of Variance  0.618 0.128 0.119 0.0667 0.0426 0.0163
Cumulative Proportion   0.618 0.746 0.865 0.9313 0.9739 0.9902
                         PC7
Standard deviation      0.26204
Proportion of Variance  0.00981
Cumulative Proportion   1.00000
```

The linear combination for the first principal component is 2

```
R> a1 <- heptathlon_pca$rotation[,1]
R> a1
```

```
 hurdles highjump     shot  run200m longjump  javelin  run800m
 -0.4504  -0.3145  -0.4025  -0.4271  -0.4510  -0.2423  -0.3029
```

We see that the hurdles and long jump events receive the highest weight but
the javelin result is less important. For computing the first principal component, the data need to be rescaled appropriately. The center and the scaling
used by prcomp internally can be extracted from the heptathlon_pca via

```
R> center <- heptathlon_pca$center
R> scale <- heptathlon_pca$scale
```

Now, we can apply the scale function to the data and multiply it with the
loadings matrix in order to compute the first principal component score for
each competitor

```
R> hm <- as.matrix(heptathlon[,-score])
R> drop(scale(hm, center = center, scale = scale) %*%
+       heptathlon_pca$rotation[,1])
```

Joyner-Kersee (USA)	John (GDR)	Behmer (GDR)
-4.757530	-3.147943	-2.926185
Sablovskaite (URS)	Choubenkova (URS)	Schulz (GDR)
-1.288136	-1.503451	-0.958467
Fleming (AUS)	Greiner (USA)	Lajbnerova (CZE)
-0.953445	-0.633239	-0.381572
Bouraga (URS)	Wijnsma (HOL)	Dimitrova (BUL)
-0.522322	-0.217701	-1.075984
Scheider (SWI)	Braun (FRG)	Ruotsalainen (FIN)
0.003015	0.109184	0.208868
Yuping (CHN)	Hagger (GB)	Brown (USA)
0.232507	0.659520	0.756855
Mulliner (GB)	Hautenauve (BEL)	Kytola (FIN)
1.880933	1.828170	2.118203
Geremias (BRA)	Hui-Ing (TAI)	Jeong-Mi (KOR)
2.770706	3.901167	3.896848

or, more conveniently, by extracting the first from all pre-computed principal components:

```
R> predict(heptathlon_pca)[,1]
```

Joyner-Kersee (USA)	John (GDR)	Behmer (GDR)
-4.757530	-3.147943	-2.926185
Sablovskaite (URS)	Choubenkova (URS)	Schulz (GDR)
-1.288136	-1.503451	-0.958467
Fleming (AUS)	Greiner (USA)	Lajbnerova (CZE)
-0.953445	-0.633239	-0.381572
Bouraga (URS)	Wijnsma (HOL)	Dimitrova (BUL)
-0.522322	-0.217701	-1.075984
Scheider (SWI)	Braun (FRG)	Ruotsalainen (FIN)
0.003015	0.109184	0.208868
Yuping (CHN)	Hagger (GB)	Brown (USA)
0.232507	0.659520	0.756855
Mulliner (GB)	Hautenauve (BEL)	Kytola (FIN)
1.880933	1.828170	2.118203
Geremias (BRA)	Hui-Ing (TAI)	Jeong-Mi (KOR)
2.770706	3.901167	3.896848

The first two components account for 75% of the variance. A barplot of each component's variance (see Figure 3.6) shows how the first two components dominate.

```
R> plot(heptathlon_pca)
```

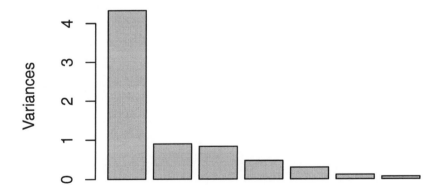

Fig. 3.6. Barplot of the variances explained by the principal components (with observations for PNG removed).

The correlation between the score given to each athlete by the standard scoring system used for the heptathlon and the first principal component score can be found from

```
R> cor(heptathlon$score, heptathlon_pca$x[,1])
```

```
[1] -0.9931
```

This implies that the first principal component is in good agreement with the score assigned to the athletes by official Olympic rules; a scatterplot of the official score and the first principal component is given in Figure 3.7. (The fact that the correlation is negative is unimportant here because of the arbitrariness of the signs of the coefficients defining the first principal component; it is the magnitude of the correlation that is important.)

```
R> plot(heptathlon$score, heptathlon_pca$x[,1])
```

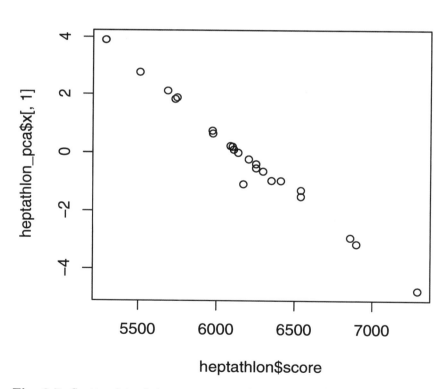

Fig. 3.7. Scatterplot of the score assigned to each athlete in 1988 and the first principal component.

3.10.3 Air pollution in US cities

In this subsection, we will return to the air pollution data introduced in Chapter 1. The data were originally collected to investigate the determinants of pollution, presumably by regressing SO2 on the six other variables. Here, however, we shall examine how principal components analysis can be used to explore various aspects of the data, and will then look at how such an analysis can also be used to address the determinants of pollution question.

To begin we shall ignore the SO2 variable and concentrate on the others, two of which relate to human ecology (popul, manu) and four to climate (temp, Wind, precip, predays). A case can be made to use negative temperature values in subsequent analyses since then all six variables are such that high values represent a less attractive environment. This is, of course, a personal view, but as we shall see later, the simple transformation of temp does aid interpretation.

Prior to undertaking the principal components analysis on the air pollution data, we will again construct a scatterplot matrix of the six variables, but here we include the histograms for each variable on the main diagonal. The diagram that results is shown in Figure 3.8.

A clear message from Figure 3.8 is that there is at least one city, and probably more than one, that should be considered an outlier. (This should come as no surprise given the investigation of the data in Chapter 2.) On the manu variable, for example, Chicago, with a value of 3344, has about twice as many manufacturing enterprises employing 20 or more workers as the city with the second highest number (Philadelphia). We shall return to this potential problem later in the chapter, but for the moment we shall carry on with a principal components analysis of the data for *all* 41 cities.

For the data in Table 1.5, it seems necessary to extract the principal components from the correlation rather than the covariance matrix, since the six variables to be used are on very different scales. The correlation matrix and the principal components of the data can be obtained in R using the following command line code:

```
R> cor(USairpollution[,-1])
```

```
              manu    popul     wind   precip predays  negtemp
manu       1.00000  0.95527  0.23795 -0.03242 0.13183  0.19004
popul      0.95527  1.00000  0.21264 -0.02612 0.04208  0.06268
wind       0.23795  0.21264  1.00000 -0.01299 0.16411  0.34974
precip    -0.03242 -0.02612 -0.01299  1.00000 0.49610 -0.38625
predays    0.13183  0.04208  0.16411  0.49610 1.00000  0.43024
negtemp    0.19004  0.06268  0.34974 -0.38625 0.43024  1.00000
```

```
R> usair_pca <- princomp(USairpollution[,-1], cor = TRUE)
```

```
R> data("USairpollution", package = "HSAUR2")
R> panel.hist <- function(x, ...) {
+       usr <- par("usr"); on.exit(par(usr))
+       par(usr = c(usr[1:2], 0, 1.5) )
+       h <- hist(x, plot = FALSE)
+       breaks <- h$breaks; nB <- length(breaks)
+       y <- h$counts; y <- y/max(y)
+       rect(breaks[-nB], 0, breaks[-1], y, col="grey", ...)
+ }
R> USairpollution$negtemp <- USairpollution$temp * (-1)
R> USairpollution$temp <- NULL
R> pairs(USairpollution[,-1], diag.panel = panel.hist,
+       pch = ".", cex = 1.5)
```

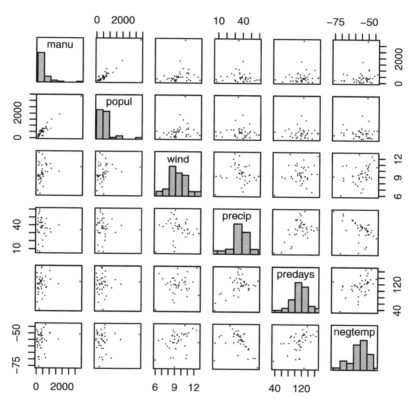

Fig. 3.8. Scatterplot matrix of six variables in the air pollution data.

R> summary(usair_pca, loadings = TRUE)

Importance of components:

	Comp.1	Comp.2	Comp.3	Comp.4	Comp.5
Standard deviation	1.482	1.225	1.1810	0.8719	0.33848
Proportion of Variance	0.366	0.250	0.2324	0.1267	0.01910
Cumulative Proportion	0.366	0.616	0.8485	0.9752	0.99426

	Comp.6
Standard deviation	0.185600
Proportion of Variance	0.005741
Cumulative Proportion	1.000000

Loadings:

	Comp.1	Comp.2	Comp.3	Comp.4	Comp.5	Comp.6
manu	−0.612	0.168	−0.273	−0.137	0.102	0.703
popul	−0.578	0.222	−0.350			−0.695
wind	−0.354	−0.131	0.297	0.869	−0.113	
precip		−0.623	−0.505	0.171	0.568	
predays	−0.238	−0.708		−0.311	−0.580	
negtemp	−0.330	−0.128	0.672	−0.306	0.558	−0.136

One thing to note about the correlations is the very high values for manu and popul, a finding we will return to later. We see that the first three components all have variances (eigenvalues) greater than one and together account for almost 85% of the variance of the original variables. Scores on these three components might be used to graph the data with little loss of information. We shall illustrate this possibility later.

Many users of principal components analysis might be tempted to search for an interpretation of the derived components that allows them to be "labelled" in some sense. This requires examining the coefficients defining each component (in the output shown above, these are scaled so that their sums of squares equal unity–"blanks" indicate near zero values). We see that the first component might be regarded as some index of "quality of life", with high values indicating a relatively poor environment (in the authors' opinion at least). The second component is largely concerned with a city's rainfall having high coefficients for precip and predays and might be labelled as the "wet weather" component. Component three is essentially a contrast between precip and negtemp and will separate cities having high temperatures and high rainfall from those that are colder but drier. A suitable label might be simply "climate type".

Attempting to label components in this way is common, but perhaps it should be a little less common; the following quotation from Marriott (1974) should act as a salutary warning about the dangers of overinterpretation:

It must be emphasised that no mathematical method is, or could be, designed to give physically meaningful results. If a mathematical expression of this sort has an obvious physical meaning, it must be attributed to a lucky change, or to the fact that the data have a strongly marked structure that shows up in analysis. Even in the latter case, quite small sampling fluctuations can upset the interpretation; for example, the first two principal components may appear in reverse order, or may become confused altogether. Reification then requires considerable skill and experience if it is to give a true picture of the physical meaning of the data.

Even if we do not care to label the three components, they can still be used as the basis of various graphical displays of the cities. In fact, this is often the most useful aspect of a principal components analysis since regarding the principal components analysis as a means of providing an informative view of multivariate data has the advantage of making it less urgent or tempting to try to interpret and label the components. The first few component scores provide a low-dimensional "map" of the observations in which the Euclidean distances between the points representing the individuals best approximate in some sense the Euclidean distances between the individuals based on the original variables. We shall return to this point in Chapter 4.

So we will begin by looking at the scatterplot matrix of the first three principal components and in each panel show the relevant bivariate boxplot; points are labelled by abbreviated city names. The plot and the code used to create it are shown in Figure 3.9.

The plot again demonstrates clearly that Chicago is an outlier and suggests that Phoenix and Philadelphia may also be suspects in this respect. Phoenix appears to offer the best quality of life (on the limited basis of the six variables recorded), and Buffalo is a city to avoid if you prefer a drier environment. We leave further interpretation to the readers.

We will now consider the main goal in the researcher's mind when collecting the air pollution data, namely determining which of the climate and human ecology variables are the best predictors of the degree of air pollution in a city as measured by the sulphur dioxide content of the air. This question would normally be addressed by multiple linear regression, but there is a potential problem with applying this technique to the air pollution data, and that is the very high correlation between the manu and popul variables. We might, of course, deal with this problem by simply dropping either manu or popul, but here we will consider a possible alternative approach, and that is regressing the SO2 levels on the principal components derived from the six other variables in the data (see Figure 3.10). Using principal components in regression analysis is discussed in detail in Jolliffe (2002); here we simply give an example. The first question we need to ask is "how many principal components should be used as explanatory variables in the regression?" The obvious answer to this question is to use the number of principal components that were identified as important in the original analysis; for example, those with eigenvalues greater than one. But this is a case where the obvious answer is

```
R> pairs(usair_pca$scores[,1:3], ylim = c(-6, 4), xlim = c(-6, 4),
+       panel = function(x,y, ...) {
+           text(x, y, abbreviate(row.names(USairpollution)),
+               cex = 0.6)
+           bvbox(cbind(x,y), add = TRUE)
+       })
```

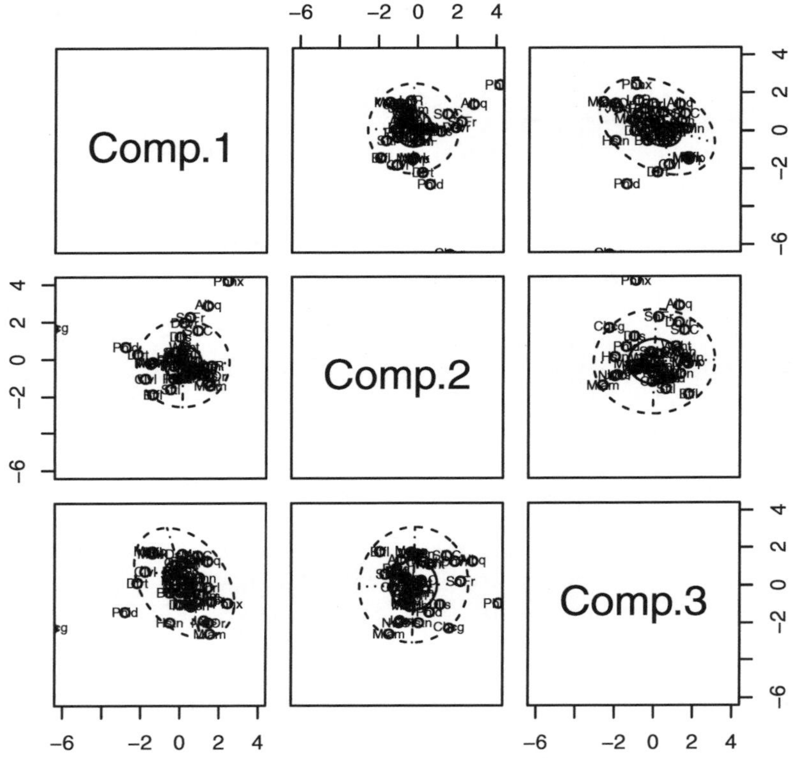

Fig. 3.9. Bivariate boxplots of the first three principal components.

not necessarily correct and, Jolliffe (2002) gives an example where there are 13 original explanatory variables and therefore 13 principal components to consider, of which only the first four have variances greater than one. But using all the principal components as explanatory variables shows that, for example, component 12, with variance 0.04, is a significant predictor of the response. So, bearing this example in mind, we will regress the SO2 variables on all six principal components; the necessary R code is given on page 92.

```
R> out <- sapply(1:6, function(i) {
+       plot(USairpollution$SO2,usair_pca$scores[,i],
+            xlab = paste("PC", i, sep = ""),
+            ylab = "Sulphur dioxide concentration")
+       })
```

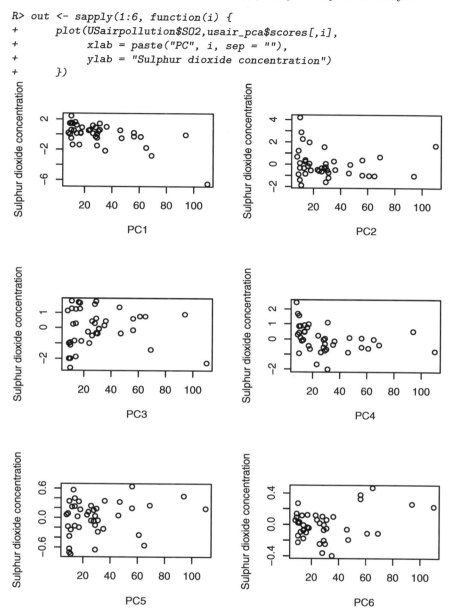

Fig. 3.10. Sulphur dioxide concentration depending on principal components.

```
R> usair_reg <- lm(SO2 ~ usair_pca$scores,
+                  data = USairpollution)
R> summary(usair_reg)

Call:
lm(formula = SO2 ~ usair_pca$scores, data = USairpollution)

Residuals:
   Min     1Q Median    3Q    Max
-23.00  -8.54  -0.99   5.76  48.76

Coefficients:
                         Estimate Std. Error t value Pr(>|t|)
(Intercept)                30.049      2.286   13.15  6.9e-15
usair_pca$scoresComp.1     -9.942      1.542   -6.45  2.3e-07
usair_pca$scoresComp.2     -2.240      1.866   -1.20   0.2384
usair_pca$scoresComp.3     -0.375      1.936   -0.19   0.8475
usair_pca$scoresComp.4     -8.549      2.622   -3.26   0.0025
usair_pca$scoresComp.5     15.176      6.753    2.25   0.0312
usair_pca$scoresComp.6     39.271     12.316    3.19   0.0031

Residual standard error: 14.6 on 34 degrees of freedom
Multiple R-squared: 0.67,        Adjusted R-squared: 0.611
F-statistic: 11.5 on 6 and 34 DF,  p-value: 5.42e-07
```

Clearly, the first principal component score is the most predictive of sulphur dioxide concentration, but it is also clear that components with small variance do not necessarily have small correlations with the response. We leave it as an exercise for the reader to investigate this example in more detail.

3.11 The biplot

A *biplot* is a graphical representation of the information in an $n \times p$ data matrix. The "bi" reflects that the technique displays in a single diagram the variances *and* covariances of the variables and the distances (see Chapter 1) between units. The technique is based on the singular value decomposition of a matrix (see Gabriel 1971).

A biplot is a two-dimensional representation of a data matrix obtained from eigenvalues and eigenvectors of the covariance matrix and obtained as

$$\mathbf{X}_2 = (\mathbf{p}_1, \mathbf{p}_2) \begin{pmatrix} \sqrt{\lambda_1} & 0 \\ 0 & \sqrt{\lambda_2} \end{pmatrix} \begin{pmatrix} \mathbf{q}_1^\top \\ \mathbf{q}_2^\top \end{pmatrix},$$

where $\mathbf{X}_2$ is the "rank two" approximation of the data matrix $\mathbf{X}$, λ_1 and λ_2 are the first two eigenvalues of the matrix $n\mathbf{S}$, and $\mathbf{q}_1$ and $\mathbf{q}_2$ are the corresponding eigenvectors. The vectors $\mathbf{p}_1$ and $\mathbf{p}_2$ are obtained as

$$\mathbf{p}_i = \frac{1}{\sqrt{\lambda_i}}\mathbf{X}\mathbf{q}_i; \quad i = 1, 2.$$

The biplot is the plot of the n rows of $\sqrt{n}(\mathbf{p}_1, \mathbf{p}_2)$ and the q rows of $n^{-1/2}(\sqrt{\lambda_1}\mathbf{q}_1, \sqrt{\lambda_2}\mathbf{q}_2)$ represented as vectors. The distance between the points representing the units reflects the generalised distance between the units (see Chapter 1), the length of the vector from the origin to the coordinates representing a particular variable reflects the variance of that variable, and the correlation of two variables is reflected by the angle between the two corresponding vectors for the two variables–the smaller the angle, the greater the correlation. Full technical details of the biplot are given in Gabriel (1981) and in Gower and Hand (1996). The biplot for the heptathlon data omitting the PNG competitor is shown in Figure 3.11. The plot in Figure 3.11 clearly shows that the winner of the gold medal, Jackie Joyner-Kersee, accumulates the majority of her points from the three events long jump, hurdles, and 200 m. We can also see from the biplot that the results of the 200 m, the hurdles and the long jump are highly correlated, as are the results of the javelin and the high jump; the 800 m time has relatively small correlation with all the other events and is almost uncorrelated with the high jump and javelin results. The first component largely separates the competitors by their overall score, with the second indicating which are their best events; for example, John, Choubenkova, and Behmer are placed near the end of the vector, representing the 800 m event because this is, relatively speaking, the event in which they give their best performance. Similarly Yuping, Scheider, and Braun can be seen to do well in the high jump. We shall have a little more to say about the biplot in the next chapter.

3.12 Sample size for principal components analysis

There have been many suggestions about the number of units needed when applying principal components analysis. Intuitively, larger values of n should lead to more convincing results and make these results more generalisable. But unfortunately many of the suggestions made, for example that n should be greater than 100 or that n should be greater than five times the number of variables, are based on minimal empirical evidence. However, Guadagnoli and Velicer (1988) review several studies that reach the conclusion that it is the minimum value of n rather than the ratio of n to q that is most relevant, although the range of values suggested for the minimum value of n in these papers, from 50 to 400, sheds some doubt on their value. And indeed other authors, for example Gorsuch (1983) and Hatcher (1994), lean towards the ratio of the minimum value of n to q as being of greater importance and recommend at least 5:1.

Perhaps the most detailed investigation of the problem is that reported in Osborne and Costello (2004), who found that the "best" results from principal

```
R> biplot(heptathlon_pca, col = c("gray", "black"))
```

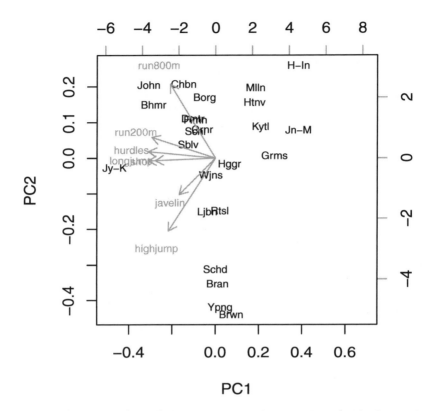

Fig. 3.11. Biplot of the (scaled) first two principal components (with observations for PNG removed).

components analysis result when n and the ratio of n to q are both large. But the actual values needed depend largely on the separation of the eigenvalues defining the principal components structure. If these eigenvalues are "close together", then a larger number of units will be needed to uncover the structure precisely than if they are far apart.

3.13 Canonical correlation analysis

Principal components analysis considers interrelationships *within* a set of variables. But there are situations where the researcher may be interested in assessing the relationships *between* two sets of variables. For example, in psychology, an investigator may measure a number of aptitude variables and a number of achievement variables on a sample of students and wish to say something about the relationship between "aptitude" and "achievement". And Krzanowski (1988) suggests an example in which an agronomist has taken, say, q_1 measurements related to the yield of plants (e.g., height, dry weight, number of leaves) at each of n sites in a region and at the same time may have recorded q_2 variables related to the weather conditions at these sites (e.g., average daily rainfall, humidity, hours of sunshine). The whole investigation thus consists of taking $(q_1 + q_2)$ measurements on n units, and the question of interest is the measurement of the association between "yield" and "weather". One technique for addressing such questions is *canonical correlation analysis*, although it has to be said at the outset that the technique is used less widely than other multivariate techniques, perhaps because the results from such an analysis are frequently difficult to interpret. For these reasons, the account given here is intentionally brief.

One way to view canonical correlation analysis is as an extension of multiple regression where a single variable (the response) is related to a number of explanatory variables and the regression solution involves finding the linear combination of the explanatory variables that is most highly correlated with the response. In canonical correlation analysis where there is more than a single variable in each of the two sets, the objective is to find the linear functions of the variables in one set that maximally correlate with linear functions of variables in the other set. Extraction of the coefficients that define the required linear functions has similarities to the process of finding principal components. A relatively brief account of the technical aspects of canonical correlation analysis (CCA) follows; full details are given in Krzanowski (1988) and Mardia et al. (1979).

The purpose of canonical correlation analysis is to characterise the independent statistical relationships that exist between two sets of variables, $\mathbf{x}^\top = (x_1, x_2, \ldots, x_{q_1})$ and $\mathbf{y}^\top = (y_1, y_2, \ldots, y_{q_2})$. The overall $(q_1 + q_2) \times (q_1 + q_2)$ correlation matrix contains all the information on associations between pairs of variables in the two sets, but attempting to extract from this matrix some idea of the association between the two sets of variables is not straightforward. This is because the correlations between the two sets may not have a consistent pattern, and these between-set correlations need to be adjusted in some way for the within-set correlations. The question of interest is "how do we quantify the association between the two sets of variables $\mathbf{x}$ and $\mathbf{y}$?" The approach adopted in CCA is to take the association between $\mathbf{x}$ and $\mathbf{y}$ to be the largest correlation between two single variables, u_1 and v_1, derived from $\mathbf{x}$ and $\mathbf{y}$, with u_1 being a linear combination of $x_1, x_2, \ldots, x_{q_1}$ and v_1 being a linear combination of $y_1, y_2, \ldots, y_{q_2}$. But often a single pair of variables (u_1, v_1) is not sufficient to quantify the association between the $\mathbf{x}$ and $\mathbf{y}$ variables, and we may need to consider some or all of s pairs $(u_1, v_1), (u_2, v_2), \ldots, (u_s, v_s)$ to do this, where $s = \min(q_1, q_2)$. Each u_i is a linear combination of the variables in $\mathbf{x}$, $u_i = \mathbf{a}_i^\top \mathbf{x}$, and each v_i is a linear combination of the variables $\mathbf{y}$,

$v_i = \mathbf{b}_i^{\top}\mathbf{y}$, with the coefficients $(\mathbf{a}_i, \mathbf{b}_i)\,(i = 1\ldots s)$ being chosen so that the u_i and v_i satisfy the following:

1. The u_i are mutually uncorrelated; i.e., $\mathsf{Cov}(u_i, u_j) = 0$ for $i \neq j$.
2. The v_i are mutually uncorrelated; i.e., $\mathsf{Cov}(v_i, v_j) = 0$ for $i \neq j$.
3. The correlation between u_i and v_i is R_i for $i = 1\ldots s$, where $R_1 > R_2 > \cdots > R_s$. The R_i are the *canonical correlations*.
4. The u_i are uncorrelated with all v_j except v_i; i.e., $\mathsf{Cov}(u_i, v_j) = 0$ for $i \neq j$.

The vectors $\mathbf{a}_i$ and $\mathbf{b}_i\ i = 1, \ldots, s$, which define the required linear combinations of the x and y variables, are found as the eigenvectors of matrices $\mathbf{E}_1(q_1 \times q_1)$ (the $\mathbf{a}_i$) and $\mathbf{E}_2(q_2 \times q_2)$ (the $\mathbf{b}_i$), defined as

$$\mathbf{E}_1 = \mathbf{R}_{11}^{-1}\mathbf{R}_{12}\mathbf{R}_{22}^{-1}\mathbf{R}_{21}, \quad \mathbf{E}_2 = \mathbf{R}_{22}^{-1}\mathbf{R}_{21}\mathbf{R}_{11}^{-1}\mathbf{R}_{12},$$

where $\mathbf{R}_{11}$ is the correlation matrix of the variables in $\mathbf{x}$, $\mathbf{R}_{22}$ is the correlation matrix of the variables in $\mathbf{y}$, and $\mathbf{R}_{12} = \mathbf{R}_{21}$ is the $q_1 \times q_2$ matrix of correlations across the two sets of variables. The canonical correlations $R_1, R_2, \ldots, R_s$ are obtained as the square roots of the non-zero eigenvalues of either $\mathbf{E}_1$ or $\mathbf{E}_2$. The s canonical correlations $R_1, R_2, \ldots, R_s$ express the association between the $\mathbf{x}$ and $\mathbf{y}$ variables after removal of the within-set correlation.

Inspection of the coefficients of each original variable in each canonical variate can provide an interpretation of the canonical variate in much the same way as interpreting principal components. Such interpretation of the canonical variates may help to describe just how the two sets of original variables are related (see Krzanowski 2010). In practise, interpretation of canonical variates can be difficult because of the possibly very different variances and covariances among the original variables in the two sets, which affects the sizes of the coefficients in the canonical variates. Unfortunately, there is no convenient normalisation to place all coefficients on an equal footing (see Krzanowski 2010). In part, this problem can be dealt with by restricting interpretation to the standardised coefficients; i.e., the coefficients that are appropriate when the original variables have been standardised.

We will now look at two relatively simple examples of the application of canonical correlation analysis.

3.13.1 Head measurements

As our first example of CCA, we shall apply the technique to data on head length and head breadth for each of the first two adult sons in 25 families shown in Table 3.1. (Part of these data were used earlier in the chapter.) These data were collected by Frets (1921), and the question that was of interest to Frets was whether there is a relationship between the head measurements for pairs of sons. We shall address this question by using canonical correlation analysis. Here we shall develop the canonical correlation analysis from first principles as detailed above. Assuming the head measurements data are contained in the data frame `headsize`, the necessary R code is

```
R> headsize.std <- sweep(headsize, 2,
+                          apply(headsize, 2, sd), FUN = "/")
R> R <- cor(headsize.std)
R> r11 <- R[1:2, 1:2]
R> r22 <- R[-(1:2), -(1:2)]
R> r12 <- R[1:2, -(1:2)]
R> r21 <- R[-(1:2), 1:2]
R> (E1 <- solve(r11) %*% r12 %*% solve(r22) %*%r21)
```

```
          head1 breadth1
head1    0.3225   0.3168
breadth1 0.3019   0.3021
```

```
R> (E2 <- solve(r22) %*% r21 %*% solve(r11) %*%r12)
```

```
          head2 breadth2
head2    0.3014   0.3002
breadth2 0.3185   0.3232
```

```
R> (e1 <- eigen(E1))
```

```
$values
[1] 0.621745 0.002888
```

```
$vectors
        [,1]     [,2]
[1,] 0.7270 -0.7040
[2,] 0.6866  0.7102
```

```
R> (e2 <- eigen(E2))
```

```
$values
[1] 0.621745 0.002888
```

```
$vectors
         [,1]     [,2]
[1,] -0.6838 -0.7091
[2,] -0.7297  0.7051
```

Here the four linear functions are found to be

$$u_1 = +0.73x_1 + 0.69x_2,$$
$$u_2 = -0.70x_1 + 0.71x_2,$$
$$v_1 = -0.68x_3 - 0.73x_4,$$
$$v_2 = -0.71x_3 + 0.71x_4.$$

```
R> girth1 <- headsize.std[,1:2] %*% e1$vectors[,1]
R> girth2 <- headsize.std[,3:4] %*% e2$vectors[,1]
```

```
R> shape1 <- headsize.std[,1:2] %*% e1$vectors[,2]
R> shape2 <- headsize.std[,3:4] %*% e2$vectors[,2]
R> (g <- cor(girth1, girth2))

        [,1]
[1,] -0.7885

R> (s <- cor(shape1, shape2))

        [,1]
[1,] 0.05374


R> plot(girth1, girth2)
R> plot(shape1, shape2)
```

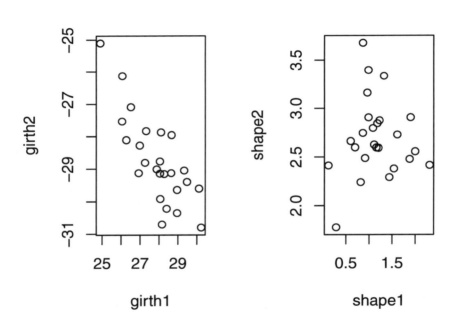

Fig. 3.12. Scatterplots of girth and shape for first and second sons.

The first canonical variate for both first and second sons is simply a weighted sum of the two head measurements and might be labelled "girth"; these two variates have a correlation of −0.79. (The negative value arises because of the arbitrariness of the sign in the first coefficient of an eigenvector— here both coefficients for girth in first sons are positive and for second sons they are both negative. The correlation can also be found as the square root of the first eigenvalue of E1 (and E2), namely 0.6217.) Each second canonical variate is a weighted difference of the two head measurements and can be interpreted roughly as head "shape"; here the correlation is 0.05 (which can also be found as the square root of the second eigenvalue of E1, namely 0.0029). (Girth and shape are defined to be uncorrelated bioth within and between first and second sons.)

In this example, it is clear that the association between the two head measurements of first and second sons is almost entirely expressed through the "girth" variables, with the two "shape" variables being almost uncorrelated. The association between the two sets of measurements is essentially one-dimensional. A scatterplot of girth for first and second sons and a similar plot for shape reinforce this conclusion. Both plots are shown in Figure 3.12.

3.13.2 Health and personality

We can now move on to a more substantial example taken from Affifi, Clark, and May (2004) and also discussed by Krzanowski (2010). The data for this example arise from a study of depression amongst 294 respondents in Los Angeles. The two sets of variables of interest were "health variables", namely the CESD (the sum of 20 separate numerical scales measuring different aspects of depression) and a measure of general health and "personal" variables, of which there were four: gender, age, income, and educational level (numerically coded from the lowest "less than high school", to the highest, "finished doctorate"). The sample correlation matrix between these variables is given in Table 3.3.

Table 3.3: LAdepr data. Los Angeles Depression Data.

CESD	Health	Gender	Age	Edu	Income
1.000	0.212	0.124	-0.164	-0.101	-0.158
0.212	1.000	0.098	0.308	-0.207	-0.183
0.124	0.098	1.000	0.044	-0.106	-0.180
-0.164	0.308	0.044	1.000	-0.208	-0.192
-0.101	-0.207	-0.106	-0.208	1.000	0.492
-0.158	-0.183	-0.180	-0.192	0.492	1.000

Here the maximum number of canonical variate pairs is two, and they can be found using the following R code:

```
R> r11 <- LAdepr[1:2, 1:2]
R> r22 <- LAdepr[-(1:2), -(1:2)]
R> r12 <- LAdepr[1:2, -(1:2)]
R> r21 <- LAdepr[-(1:2), 1:2]
R> (E1 <- solve(r11) %*% r12 %*% solve(r22) %*%r21)

           CESD    Health
CESD    0.08356 -0.04312
Health -0.03256  0.13338

R> (E2 <- solve(r22) %*% r21 %*% solve(r11) %*%r12)

           Gender       Age      Edu    Income
Gender   0.015478 -0.001483 -0.01813 -0.02224
Age     -0.002445  0.147069 -0.03981 -0.01599
Edu     -0.014915 -0.026019  0.02544  0.02508
Income  -0.021163  0.013027  0.02159  0.02895

R> (e1 <- eigen(E1))

$values
[1] 0.15347 0.06347

$vectors
          [,1]    [,2]
[1,]   0.5250 -0.9065
[2,]  -0.8511 -0.4222

R> (e2 <- eigen(E2))

$values
[1]  1.535e-01  6.347e-02 -8.760e-18  6.633e-19

$vectors
             [,1]     [,2]     [,3]     [,4]
[1,]  -0.002607  0.4904  -0.4617  0.79705
[2,]  -0.980095 -0.3208  -0.1865  0.06251
[3,]   0.185801 -0.4270  -0.7974 -0.05817
[4,]  -0.069886 -0.6887   0.3409  0.59784
```

(Note that the third and fourth eigenvalues of E2 are essentially zero, as we would expect in this case.) The first canonical correlation is 0.409, calculated as the square root of the first eigenvalue of E1, which is given above as 0.15347. If tested as outlined in Exercise 3.4, it has an associated p-value that is very small; there is strong evidence that the first canonical correlation is significant. The corresponding variates, in terms of standardised original variables, are

$$u_1 = +0.53 \text{ CESD} - 0.85 \text{ Health}$$
$$v_1 = -0.00 \text{ Gender} - 0.98 \text{ Age} + 0.19 \text{ Education} - 0.07 \text{ Income} .$$

High coefficients correspond to CESD (positively) and health (negatively) for the perceived health variables, and to Age (negatively) and Education (positively) for the personal variables. It appears that relatively older and medicated people tend to have lower depression scores, but perceive their health as relatively poor, while relatively younger but educated people have the opposite health perception. (We are grateful to Krzanowski 2010, for this interpretation.)

The second canonical correlation is 0.261, calculated as the square root of the second eigenvalue, found from the R output above to be 0.06347; this correlation is also significant (see Exercise 3.4). The corresponding canonical variates are

$$u_1 = -0.91 \text{ CESD } - 0.42 \text{ Health}$$
$$v_1 = +0.49 \text{ Gender } - 0.32 \text{ Age } - 0.43 \text{ Education } - 0.69 \text{ Income }.$$

Since the higher value of the Gender variable is for females, the interpretation here is that relatively young, poor, and uneducated females are associated with higher depression scores and, to a lesser extent, with poor perceived health (again this interpretation is due to Krzanowski 2010).

3.14 Summary

Principal components analysis is among the oldest of multivariate techniques, having been introduced originally by Pearson (1901) and independently by Hotelling (1933). It remains, however, one of the most widely employed methods of multivariate analysis, useful both for providing a convenient method of displaying multivariate data in a lower dimensional space and for possibly simplifying other analyses of the data. The reduction in dimensionality that can often be achieved by a principal components analysis is possible only if the original variables are correlated; if the original variables are independent of one another a principal components analysis cannot lead to any simplification.

The examples of principal components analysis given in this chapter have involved variables that are continuous. But when used as a descriptive technique there is no reason for the variables in the analysis to be of any particular type. So variables might be a mixture of continuous, ordinal, and even binary variables. Linear functions of continuous variables are perhaps easier to interpret than corresponding functions of binary variables, but the basic objective of principal components analysis, namely to summarise most of the "variation" present in the original q variables, can be achieved regardless of the nature of the observed variables.

The results of a canonical correlation analysis have the reputation of often being difficult to interpret, and in many respects this is a reputation that is well earned. Certainly one has to know the variables very well to have any hope of extracting a convincing explanation of the results. But in some circumstances (the heads measurement data is an example) CCA can provide a useful description and informative insights into the association between two sets of variables.

Finally, we should mention a technique known as independent component analysis (ICA), a potentially powerful technique that seeks to uncover hidden variables in high-dimensional data. In particular, ICA is often able to separate independent sources linearly mixed in several sensors; for example, in prenatal diagnostics a multichannel electrocardiogram (ECG) may be used to record a mixture of maternal and foetal electrical activity, including foetal heart rate and maternal heart rate. However, the maternal ECG will be much stronger than the foetal signal, and the signal recorded is also likely to be affected by other sources of electrical interference. Such data can be separated into their component signals by ICA, as demonstrated in Izenman (2008).

Similar mixtures of signals arise in many other areas, for example monitoring of human brain-wave activity and functional magnetic resonance imaging experiments, and consequently ICA has become increasingly used in such areas. Details of the methods are given in Hyvärinen (1999) and Hyvärinen, Karhunen, and Oja (2001). Interestingly, the first step in analysing multivariate data with ICA is often a principal components analysis. We have chosen not to describe ICA in this book because it is technically demanding and also rather specialised, but we should mention in passing that the method is available in R via the add-on package **fastICA** (Marchini, Heaton, and Ripley 2010).

3.15 Exercises

Ex. 3.1 Construct the scatterplot of the heptathlon data showing the contours of the estimated bivariate density function on each panel. Is this graphic more useful than the unenhanced scatterplot matrix?

Ex. 3.2 Construct a diagram that shows the SO2 variable in the air pollution data plotted against each of the six explanatory variables, and in each of the scatterplots show the fitted linear regression and a fitted locally weighted regression. Does this diagram help in deciding on the most appropriate model for determining the variables most predictive of sulphur dioxide levels?

Ex. 3.3 Find the principal components of the following correlation matrix given by MacDonnell (1902) from measurements of seven physical characteristics in each of 3000 convicted criminals:

$$R = \begin{matrix} \text{Head length} \\ \text{Head breadth} \\ \text{Face breadth} \\ \text{Left finger length} \\ \text{Left forearm length} \\ \text{Left foot length} \\ \text{Height} \end{matrix} \begin{pmatrix} 1.000 & & & & & & \\ 0.402 & 1.000 & & & & & \\ 0.396 & 0.618 & 1.000 & & & & \\ 0.301 & 0.150 & 0.321 & 1.000 & & & \\ 0.305 & 0.135 & 0.289 & 0.846 & 1.000 & & \\ 0.339 & 0.206 & 0.363 & 0.759 & 0.797 & 1.000 & \\ 0.340 & 0.183 & 0.345 & 0.661 & 0.800 & 0.736 & 1.000 \end{pmatrix}.$$

How would you interpret the derived components?

Ex. 3.4 Not all canonical correlations may be statistically significant. An approximate test proposed by Bartlett (1947) can be used to determine how many significant relationships exist. The test statistic for testing that at least one canonical correlation is significant is $\Phi_0^2 = -\{n - \frac{1}{2}(q_1 + q_2 + 1)\} \sum_{i=1}^{s} \log(1 - \lambda_i)$, where the λ_i are the eigenvalues of $\mathbf{E}_1$ and $\mathbf{E}_2$. Under the null hypothesis that all correlations are zero, Φ_0^2 has a chi-square distribution with $q_1 \times q_2$ degrees of freedom. Write R code to apply this test to the **headsize** data (Table 3.1) and the **depression** data (Table 3.3).

Ex. 3.5 Repeat the regression analysis for the air pollution data described in the text after removing whatever cities you think should be regarded as outliers. For the results given in the text and the results from the outliers-removed data, produce scatterplots of sulphur dioxide concentration against each of the principal component scores. Interpret your results.

4

Multidimensional Scaling

4.1 Introduction

In Chapter 3, we noted in passing that one of the most useful ways of using principal components analysis was to obtain a low-dimensional "map" of the data that preserved as far as possible the Euclidean distances between the observations in the space of the original q variables. In this chapter, we will make this aspect of principal component analysis more explicit and also introduce a class of other methods, labelled *multidimensional scaling*, that aim to produce similar maps of data but do not operate directly on the usual multivariate data matrix, $\mathbf{X}$. Instead they are applied to distance matrices (see Chapter 1), which are derived from the matrix $\mathbf{X}$ (an example of a distance matrix derived from a small set of multivariate data is shown in Subsection 4.4.2), and also to so-called *dissimilarity* or *similarity matrices* that arise directly in a number of ways, in particular from judgements made by human raters about how alike pairs of objects, stimuli, etc., of interest are. An example of a directly observed dissimilarity matrix is shown in Table 4.5, with judgements about political and war leaders that had major roles in World War II being given by a subject after receiving the simple instructions to rate each pair of politicians on a nine-point scale, with 1 indicating two politicians they regard as very similar and 9 indicating two they regard as very dissimilar. (If the nine point-scale had been 1 for very dissimilar and 9 for very similar, then the result would have been a rating of similarity, although similarities are often scaled to lie in a $[0, 1]$ interval. The term *proximity* is often used to encompass both dissimilarity and similarity ratings.)

4.2 Models for proximity data

Models are fitted to proximities in order to clarify, display, and help understand and possibly explain any structure or pattern amongst the observed or calculated proximities not readily apparent in the collection of numerical values. In some areas, particularly psychology, the ultimate goal in the analysis of a set of proximities is more specific, namely the development of theories for explaining similarity judgements; in other words, trying to answer the question, "what makes things seem alike or seem different?" According to Carroll and Arabie (1980) and Carroll, Clark, and DeSarbo (1984), models for the analysis of proximity data can be categorised into one of three major classes: *spatial models*, *tree models*, and *hybrid models*. In this chapter and in this book, we only deal with the first of these three classes. For details of tree models and hybrid models, see, for example, Everitt and Rabe-Hesketh (1997).

4.3 Spatial models for proximities: Multidimensional scaling (MDS)

A spatial representation of a proximity matrix consists of a set of n m-dimensional coordinates, each one of which represents one of the n units in the data. The required coordinates are generally found by minimising some measure of "fit" between the distances implied by the coordinates and the observed proximities. In simple terms, a geometrical model is sought in which the larger the observed distance or dissimilarity between two units (or the smaller their similarity), the further apart should be the points representing them in the model. In general (but not exclusively), the distances between the points in the spatial model are assumed to be Euclidean. Finding the best-fitting set of coordinates and the appropriate value of m needed to adequately represent the observed proximities is the aim of the many methods of multidimensional scaling that have been proposed. The hope is that the number of dimensions, m, will be small, ideally two or three, so that the derived spatial configuration can be easily plotted. Multidimensional scaling is essentially a data reduction technique because the aim is to find a set of points in low dimension that approximate the possibly high-dimensional configuration represented by the original proximity matrix. The variety of methods that have been proposed largely differ in how agreement between fitted distances and observed proximities is assessed. In this chapter, we will consider two methods, *classical multidimensional scaling* and *non-metric multidimensional scaling*.

4.4 Classical multidimensional scaling

First, like all MDS techniques, classical scaling seeks to represent a proximity matrix by a simple geometrical model or map. Such a model is characterised by a set of points $\mathbf{x}_1, \mathbf{x}_2, \ldots, \mathbf{x}_n$, in m dimensions, each point representing one of the units of interest, and a measure of the distance between pairs of points. The objective of MDS is to determine both the dimensionality, m, of the model, and the n m-dimensional coordinates, $\mathbf{x}_1, \mathbf{x}_2, \ldots, \mathbf{x}_n$, so that the model gives a "good" fit for the observed proximities. Fit will often be judged by some numerical index that measures how well the proximities and the distances in the geometrical model match. In essence, this simply means that the larger an observed dissimilarity between two stimuli (or the smaller their similarity), the further apart should be the points representing them in the final geometrical model.

The question now arises as to how we estimate m, and the coordinate values $x_1, x_2, \ldots, x_n$, from the observed proximity matrix. Classical scaling provides an answer to this question based on the work of Young and Householder (1938). To begin, we must note that there is no unique set of coordinate values that give rise to a set of distances since the distances are unchanged by shifting the whole configuration of points from one place to another or by rotation or reflection of the configuration. In other words, we cannot uniquely determine either the location or the orientation of the configuration. The location problem is usually overcome by placing the mean vector of the configuration at the origin. The orientation problem means that any configuration derived can be subjected to an arbitrary orthogonal transformation (see Chapter 5). Such transformations can often be used to facilitate the interpretation of solutions, as will be seen later (and again see Chapter 5).

4.4.1 Classical multidimensional scaling: Technical details

To begin our account of the method, we shall assume that the proximity matrix we are dealing with is a matrix of Euclidean distances, $\mathbf{D}$, derived from a raw $n \times q$ data matrix, $\mathbf{X}$. In Chapter 1, we saw how to calculate Euclidean distances from $\mathbf{X}$; classical multidimensional scaling is essentially concerned with the reverse problem: given the distances, how do we find $\mathbf{X}$? First assume $\mathbf{X}$ is known and consider the $n \times n$ *inner products matrix*, $\mathbf{B}$

$$\mathbf{B} = \mathbf{X}\mathbf{X}^{\top}. \tag{4.1}$$

The elements of $\mathbf{B}$ are given by

$$b_{ij} = \sum_{k=1}^{q} x_{ik}x_{jk}. \tag{4.2}$$

It is easy to see that the squared Euclidean distances between the rows of $\mathbf{X}$ can be written in terms of the elements of $\mathbf{B}$ as

$$d_{ij}^2 = b_{ii} + b_{jj} - 2b_{ij}. \tag{4.3}$$

If the bs could be found in terms of the ds in the equation above, then the required coordinate values could be derived by factoring $\mathbf{B}$ as in (4.1). No unique solution exists unless a location constraint is introduced; usually the centre of the points $\overline{\mathbf{x}}$ is set at the origin, so that $\sum_{i=1}^{n} x_{ik} = 0$ for all $k = 1, 2 \ldots m$. These constraints and the relationship given in (4.2) imply that the sum of the terms in any row of $\mathbf{B}$ must be zero. Consequently, summing the relationship given in (4.2) over i, over j, and finally over both i and j leads to the series of equations

$$\sum_{i=1}^{n} d_{ij}^2 = T + nb_{jj},$$

$$\sum_{j=1}^{n} d_{ij}^2 = T + nb_{ii},$$

$$\sum_{i=1}^{n} \sum_{j=1}^{n} d_{ij}^2 = 2nT,$$

where $T = \sum_{i=1}^{n} b_{ii}$ is the trace of the matrix $\mathbf{B}$. The elements of $\mathbf{B}$ can now be found in terms of squared Euclidean distances as

$$b_{ij} = -\frac{1}{2} \left(d_{ij}^2 - d_{i.}^2 - d_{.j}^2 + d_{..}^2 \right),$$

where

$$d_{i.}^2 = \frac{1}{n} \sum_{j=1}^{n} d_{ij}^2,$$

$$d_{.j}^2 = \frac{1}{n} \sum_{i=1}^{n} d_{ij}^2,$$

$$d_{..}^2 = \frac{1}{n^2} \sum_{i=1}^{n} \sum_{j=1}^{n} d_{ij}^2.$$

Having now derived the elements of $\mathbf{B}$ in terms of Euclidean distances, it remains to factor it to give the coordinate values. In terms of its spectral decomposition (see Chapter 3), $\mathbf{B}$ can be written as

$$\mathbf{B} = \mathbf{V} \boldsymbol{\Lambda} \mathbf{V}^{\top},$$

where $\boldsymbol{\Lambda} = \mathrm{diag}\,(\lambda_1, \ldots, \lambda_n)$ is the diagonal matrix of eigenvalues of $\mathbf{B}$ and $\mathbf{V} = (\mathbf{V}_1, \ldots, \mathbf{V}_n)$ the corresponding matrix of eigenvectors, normalised so that the sum of squares of their elements is unity, that is, $\mathbf{V}_i \mathbf{V}_i^{\top} = 1$. The eigenvalues are assumed to be labelled such that $\lambda_1 \geq \lambda_2 \geq \cdots \geq \lambda_n$. When $\mathbf{D}$ arises from an $n \times q$ matrix of full rank, then the rank of $\mathbf{B}$ is q, so that the last $n - q$ of its eigenvalues will be zero. So $\mathbf{B}$ can be written as

$$\mathbf{B} = \mathbf{V}_1 \boldsymbol{\Lambda}_1 \mathbf{V}_1^{\top},$$

where $\mathbf{V}_1$ contains the first q eigenvectors and $\boldsymbol{\Lambda}_1$ the q non-zero eigenvalues.

The required coordinate values are thus

$$\mathbf{X} = \mathbf{V}_1 \boldsymbol{\Lambda}_1^{\frac{1}{2}},$$

where $\boldsymbol{\Lambda}_1^{\frac{1}{2}} = \text{diag}\left(\lambda_1^{\frac{1}{2}}, \ldots, \lambda_q^{\frac{1}{2}}\right)$.

Using all q-dimensions will lead to complete recovery of the original Euclidean distance matrix. The best-fitting m-dimensional representation is given by the m eigenvectors of $\mathbf{B}$ corresponding to the m largest eigenvalues. The adequacy of the m-dimensional representation can be judged by the size of the criterion

$$P_m = \frac{\sum_{i=1}^{m} \lambda_i}{\sum_{i=1}^{n} \lambda_i}.$$

Values of P_m of the order of 0.8 suggest a reasonable fit.

It should be mentioned here that where the proximity matrix contains Euclidean distances calculated from an $n \times q$ data matrix $\mathbf{X}$, classical scaling can be shown to be equivalent to principal components analysis, with the required coordinate values corresponding to the scores on the principal component extracted from the covariance matrix of the data. One result of this duality is that classical multidimensional scaling is also referred to as *principal coordinates*–see Gower (1966). And the m-dimensional principal components solution $(m < q)$ is "best" in the sense that it minimises the measure of fit

$$S = \sum_{i=1}^{n} \sum_{i=1}^{n} \left(d_{ij}^2 - \left(d_{ij}^{(m)}\right)^2\right),$$

where d_{ij} is the Euclidean distance between individuals i and j based on their original q variable values and $d_{ij}^{(m)}$ is the corresponding distance calculated from the m principal component scores.

When the observed proximity matrix is not Euclidean, the matrix $\mathbf{B}$ is not positive-definite. In such cases, some of the eigenvalues of $\mathbf{B}$ will be negative; correspondingly, some coordinate values will be complex numbers. If, however, $\mathbf{B}$ has only a small number of small negative eigenvalues, a useful representation of the proximity matrix may still be possible using the eigenvectors associated with the m largest positive eigenvalues. The adequacy of the resulting solution might be assessed using one of the following two criteria suggested by Mardia et al. (1979):

$$P_m^{(1)} = \frac{\sum_{i=1}^{m} |\lambda_i|}{\sum_{i=1}^{n} |\lambda_i|},$$

$$P_m^{(2)} = \frac{\sum_{i=1}^{m} \lambda_i^2}{\sum_{i=1}^{n} \lambda_i^2}.$$

Again we would look for values above 0.8 to claim a "good" fit. Alternatively, Sibson (1979) recommends one of the following two criteria for deciding on the number of dimensions for the spatial model to adequately represent the observed proximities:

Trace criterion: Choose the number of coordinates so that the sum of the *positive* eigenvalues is approximately equal to the sum of *all* the eigenvalues.

Magnitude criterion: Accept as genuinely positive only those eigenvalues whose magnitude substantially exceeds that of the largest negative eigenvalue.

If, however, the matrix **B** has a considerable number of large negative eigenvalues, classical scaling of the proximity matrix may be inadvisable and some other methods of scaling, for example non-metric scaling (see the next section), might be better employed.

4.4.2 Examples of classical multidimensional scaling

For our first example we will use the small set of multivariate data **X**

```
       [,1] [,2] [,3] [,4] [,5]
 [1,]    3    4    4    6    1
 [2,]    5    1    1    7    3
 [3,]    6    2    0    2    6
 [4,]    1    1    1    0    3
 [5,]    4    7    3    6    2
 [6,]    2    2    5    1    0
 [7,]    0    4    1    1    1
 [8,]    0    6    4    3    5
 [9,]    7    6    5    1    4
[10,]    2    1    4    3    1
```

and the associated matrix of Euclidean distances (computed via the dist() function) will be our proximity matrix

```
R> (D <- dist(X))
```

```
        1     2     3     4     5     6     7     8     9
2   5.196
3   8.367 6.083
4   7.874 8.062 6.325
5   3.464 6.557 8.367 9.274
6   5.657 8.426 8.832 5.292 7.874
7   6.557 8.602 8.185 3.873 7.416 5.000
8   6.164 8.888 8.367 6.928 6.000 7.071 5.745
9   7.416 9.055 6.856 8.888 6.557 7.550 8.832 7.416
10  4.359 6.164 7.681 4.796 7.141 2.646 5.099 6.708 8.000
```

To apply classical scaling to this matrix in R, we can use the cmdscale() function to do the scaling:

```
R> cmdscale(D, k = 9, eig = TRUE)
```

```
$points
           [,1]      [,2]     [,3]     [,4]      [,5]       [,6]
 [1,]  -1.6038  -2.38061  -2.2301  -0.3657   0.11536   0.000e+00
 [2,]  -2.8246   2.30937  -3.9524   0.3419   0.33169  -2.797e-08
 [3,]  -1.6908   5.13970   1.2880   0.6503  -0.05134  -1.611e-09
 [4,]   3.9528   2.43234   0.3834   0.6864  -0.03461  -7.393e-09
 [5,]  -3.5985  -2.75538  -0.2551   1.0784  -1.26125  -5.198e-09
 [6,]   2.9520  -1.35475  -0.1899  -2.8211   0.12386  -2.329e-08
 [7,]   3.4690  -0.76411   0.3017   1.6369  -1.94210  -1.452e-08
 [8,]   0.3545  -2.31409   2.2162   2.9240   2.00450  -1.562e-08
 [9,]  -2.9362   0.01280   4.3117  -2.5123  -0.18912  -1.404e-08
[10,]   1.9257  -0.32527  -1.8734  -1.6189   0.90299   6.339e-09
           [,7] [,8] [,9]
 [1,]   1.791e-08  NaN  NaN
 [2,]  -1.209e-09  NaN  NaN
 [3,]   1.072e-09  NaN  NaN
 [4,]   1.088e-08  NaN  NaN
 [5,]  -2.798e-09  NaN  NaN
 [6,]  -7.146e-09  NaN  NaN
 [7,]   3.072e-09  NaN  NaN
 [8,]   2.589e-10  NaN  NaN
 [9,]   7.476e-09  NaN  NaN
[10,]   3.303e-09  NaN  NaN


$eig
 [1]   7.519e+01   5.881e+01   4.961e+01   3.043e+01   1.037e+01
 [6]   2.101e-15   5.769e-16  -2.819e-15  -3.233e-15  -6.274e-15

$x
NULL

$ac
[1]  0

$GOF
[1]  1 1
```

Note that as $q = 5$ in this example, eigenvalues six to nine are essentially zero and only the first five columns of `points` represent the Euclidean distance matrix. First we should confirm that the five-dimensional solution achieves complete recovery of the observed distance matrix. We can do this simply by comparing the original distances with those calculated from the five-dimensional scaling solution coordinates using the following R code:

```
R> max(abs(dist(X) - dist(cmdscale(D, k = 5))))
[1] 1.243e-14
```

This confirms that all the differences are essentially zero and that therefore the observed distance matrix is recovered by the five-dimensional classical scaling solution.

We can also check the duality of classical scaling of Euclidean distances and principal components analysis mentioned previously in the chapter by comparing the coordinates of the five-dimensional scaling solution given above with the first five principal component (up to signs) scores obtained by applying PCA to the covariance matrix of the original data; the necessary R code is

```
R> max(abs(prcomp(X)$x) - abs(cmdscale(D, k = 5)))
```

```
[1] 3.035e-14
```

Now let us look at two examples involving distances that are not Euclidean. First, we will calculate the *Manhattan distances* between the rows of the small data matrix X. The Manhattan distance for units i and j is given by $\sum_{k=1}^{q} |x_{ik} - x_{jk}|$, and these distances are not Euclidean. (Manhattan distances will be familiar to those readers who have walked around New York.) The R code for calculating the Manhattan distances and then applying classical multidimensional scaling to the resulting distance matrix is:

```
R> X_m <- cmdscale(dist(X, method = "manhattan"),
+                  k = nrow(X) - 1, eig = TRUE)
```

The criteria $P_m^{(1)}$ and $P_m^{(2)}$ can be computed from the eigenvalues as follows:

```
R> (X_eigen <- X_m$eig)
```

```
[1]   2.807e+02   2.494e+02   2.289e+02   9.251e+01   4.251e+01
[6]   2.197e+01  -7.105e-15  -1.507e+01  -2.805e+01  -5.683e+01
```

Note that some of the eigenvalues are negative in this case.

```
R> cumsum(abs(X_eigen)) / sum(abs(X_eigen))
```

```
[1] 0.2763 0.5218 0.7471 0.8382 0.8800 0.9016 0.9016 0.9165
[9] 0.9441 1.0000
```

```
R> cumsum(X_eigen^2) / sum(X_eigen^2)
```

```
[1] 0.3779 0.6764 0.9276 0.9687 0.9773 0.9796 0.9796 0.9807
[9] 0.9845 1.0000
```

The values of both criteria suggest that a three-dimensional solution seems to fit well.

Table 4.1: `airdist` data. Airline distances between ten US cities.

	ATL	ORD	DEN	HOU	LAX	MIA	JFK	SFO	SEA	IAD
ATL	0									
ORD	587	0								
DEN	1212	920	0							
HOU	701	940	879	0						
LAX	1936	1745	831	1374	0					
MIA	604	1188	1726	968	2339	0				
JFK	748	713	1631	1420	2451	1092	0			
SFO	2139	1858	949	1645	347	2594	2571	0		
SEA	218	1737	1021	1891	959	2734	2408	678	0	
IAD	543	597	1494	1220	2300	923	205	2442	2329	0

For our second example of applying classical multidimensional scaling to non-Euclidean distances, we shall use the airline distances between ten US cities given in Table 4.1. These distances are not Euclidean since they relate essentially to journeys along the surface of a sphere. To apply classical scaling to these distances and to see the eigenvalues, we can use the following R code:

```
R> airline_mds <- cmdscale(airdist, k = 9, eig = TRUE)
R> airline_mds$points
```

```
         [,1]     [,2]     [,3]      [,4]        [,5] [,6] [,7] [,8]
ATL   -434.8   724.22   440.93    0.18579  -1.258e-02  NaN  NaN  NaN
ORD   -412.6    55.04  -370.93    4.39608   1.268e+01  NaN  NaN  NaN
DEN    468.2  -180.66  -213.57   30.40857  -9.585e+00  NaN  NaN  NaN
HOU   -175.6  -515.22   362.84    9.48713  -4.860e+00  NaN  NaN  NaN
LAX   1206.7  -465.64    56.53    1.34144   6.809e+00  NaN  NaN  NaN
MIA  -1161.7  -477.98   479.60  -13.79783   2.278e+00  NaN  NaN  NaN
JFK  -1115.6   199.79  -429.67  -29.39693  -7.137e+00  NaN  NaN  NaN
SFO   1422.7  -308.66  -205.52  -26.06310  -1.983e+00  NaN  NaN  NaN
SEA   1221.5   887.20   170.45   -0.06999  -8.943e-05  NaN  NaN  NaN
IAD  -1018.9    81.90  -290.65   23.50884   1.816e+00  NaN  NaN  NaN
```

(The nineth column containing NaNs is omitted from the output.) The eigenvalues are

```
R> (lam <- airline_mds$eig)
```

```
[1]  9.214e+06  2.200e+06  1.083e+06  3.322e+03  3.859e+02
[6] -5.204e-09 -9.323e+01 -2.169e+03 -9.091e+03 -1.723e+06
```

As expected (as the distances are not Euclidean), some of the eigenvalues are negative and so we will again use the criteria $P_m^{(1)}$ and $P_m^{(2)}$ to assess how many coordinates we need to adequately represent the observed distance matrix. The values of the two criteria calculated from the eigenvalues are

```
R> cumsum(abs(lam)) / sum(abs(lam))
```

```
[1]  0.6473 0.8018 0.8779 0.8781 0.8782 0.8782 0.8782 0.8783
[9]  0.8790 1.0000
```

```
R> cumsum(lam^2) / sum(lam^2)
```

```
[1]  0.9043 0.9559 0.9684 0.9684 0.9684 0.9684 0.9684 0.9684
[9]  0.9684 1.0000
```

These values suggest that the first two coordinates will give an adequate representation of the observed distances. The scatterplot of the two-dimensional coordinate values is shown in Figure 4.1. In this two-dimensional representation, the geographical location of the cities has been very well recovered by the two-dimensional multidimensional scaling solution obtained from the airline distances.

Our next example of the use of classical multidimensional scaling will involve the data shown in Table 4.2. These data show four measurements on male Egyptian skulls from five epochs. The measurements are:

mb: maximum breadth of the skull;
bh: basibregmatic height of the skull;
bl: basialiveolar length of the skull; and
nh: nasal height of the skull.

Table 4.2: skulls data. Measurements of four variables taken from Egyptian skulls of five periods.

epoch	mb	bh	bl	nh	epoch	mb	bh	bl	nh	epoch	mb	bh	bl	nh
c4000BC	131	138	89	49	c3300BC	137	136	106	49	c200BC	132	133	90	53
c4000BC	125	131	92	48	c3300BC	126	131	100	48	c200BC	134	134	97	54
c4000BC	131	132	99	50	c3300BC	135	136	97	52	c200BC	135	135	99	50
c4000BC	119	132	96	44	c3300BC	129	126	91	50	c200BC	133	136	95	52
c4000BC	136	143	100	54	c3300BC	134	139	101	49	c200BC	136	130	99	55
c4000BC	138	137	89	56	c3300BC	131	134	90	53	c200BC	134	137	93	52
c4000BC	139	130	108	48	c3300BC	132	130	104	50	c200BC	131	141	99	55
c4000BC	125	136	93	48	c3300BC	130	132	93	52	c200BC	129	135	95	47
c4000BC	131	134	102	51	c3300BC	135	132	98	54	c200BC	136	128	93	54
c4000BC	134	134	99	51	c3300BC	130	128	101	51	c200BC	131	125	88	48
c4000BC	129	138	95	50	c1850BC	137	141	96	52	c200BC	139	130	94	53
c4000BC	134	121	95	53	c1850BC	129	133	93	47	c200BC	144	124	86	50
c4000BC	126	129	109	51	c1850BC	132	138	87	48	c200BC	141	131	97	53
c4000BC	132	136	100	50	c1850BC	130	134	106	50	c200BC	130	131	98	53
c4000BC	141	140	100	51	c1850BC	134	134	96	45	c200BC	133	128	92	51
c4000BC	131	134	97	54	c1850BC	140	133	98	50	c200BC	138	126	97	54
c4000BC	135	137	103	50	c1850BC	138	138	95	47	c200BC	131	142	95	53

Table 4.2: `skulls` data (continued).

epoch	mb	bh	bl	nh	epoch	mb	bh	bl	nh	epoch	mb	bh	bl	nh
c4000BC	132	133	93	53	c1850BC	136	145	99	55	c200BC	136	138	94	55
c4000BC	139	136	96	50	c1850BC	136	131	92	46	c200BC	132	136	92	52
c4000BC	132	131	101	49	c1850BC	126	136	95	56	c200BC	135	130	100	51
c4000BC	126	133	102	51	c1850BC	137	129	100	53	cAD150	137	123	91	50
c4000BC	135	135	103	47	c1850BC	137	139	97	50	cAD150	136	131	95	49
c4000BC	134	124	93	53	c1850BC	136	126	101	50	cAD150	128	126	91	57
c4000BC	128	134	103	50	c1850BC	137	133	90	49	cAD150	130	134	92	52
c4000BC	130	130	104	49	c1850BC	129	142	104	47	cAD150	138	127	86	47
c4000BC	138	135	100	55	c1850BC	135	138	102	55	cAD150	126	138	101	52
c4000BC	128	132	93	53	c1850BC	129	135	92	50	cAD150	136	138	97	58
c4000BC	127	129	106	48	c1850BC	134	125	90	60	cAD150	126	126	92	45
c4000BC	131	136	114	54	c1850BC	138	134	96	51	cAD150	132	132	99	55
c4000BC	124	138	101	46	c1850BC	136	135	94	53	cAD150	139	135	92	54
c3300BC	124	138	101	48	c1850BC	132	130	91	52	cAD150	143	120	95	51
c3300BC	133	134	97	48	c1850BC	133	131	100	50	cAD150	141	136	101	54
c3300BC	138	134	98	45	c1850BC	138	137	94	51	cAD150	135	135	95	56
c3300BC	148	129	104	51	c1850BC	130	127	99	45	cAD150	137	134	93	53
c3300BC	126	124	95	45	c1850BC	136	133	91	49	cAD150	142	135	96	52
c3300BC	135	136	98	52	c1850BC	134	123	95	52	cAD150	139	134	95	47
c3300BC	132	145	100	54	c1850BC	136	137	101	54	cAD150	138	125	99	51
c3300BC	133	130	102	48	c1850BC	133	131	96	49	cAD150	137	135	96	54
c3300BC	131	134	96	50	c1850BC	138	133	100	55	cAD150	133	125	92	50
c3300BC	133	125	94	46	c1850BC	138	133	91	46	cAD150	145	129	89	47
c3300BC	133	136	103	53	c200BC	137	134	107	54	cAD150	138	136	92	46
c3300BC	131	139	98	51	c200BC	141	128	95	53	cAD150	131	129	97	44
c3300BC	131	136	99	56	c200BC	141	130	87	49	cAD150	143	126	88	54
c3300BC	138	134	98	49	c200BC	135	131	99	51	cAD150	134	124	91	55
c3300BC	130	136	104	53	c200BC	133	120	91	46	cAD150	132	127	97	52
c3300BC	131	128	98	45	c200BC	131	135	90	50	cAD150	137	125	85	57
c3300BC	138	129	107	53	c200BC	140	137	94	60	cAD150	129	128	81	52
c3300BC	123	131	101	51	c200BC	139	130	90	48	cAD150	140	135	103	48
c3300BC	130	129	105	47	c200BC	140	134	90	51	cAD150	147	129	87	48
c3300BC	134	130	93	54	c200BC	138	140	100	52	cAD150	136	133	97	51

We shall calculate *Mahalanobis distances* between each pair of epochs using the `mahalanobis()` function and apply classical scaling to the resulting distance matrix. In this calculation, we shall use the estimate of the assumed common covariance matrix $\mathbf{S}$

$$\mathbf{S} = \frac{29\mathbf{S}_1 + 29\mathbf{S}_2 + 29\mathbf{S}_3 + 29\mathbf{S}_4 + 29\mathbf{S}_5}{149},$$

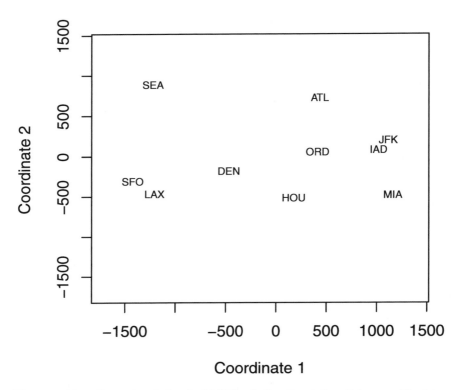

Fig. 4.1. Two-dimensional classical MDS solution for airline distances. The known spatial arrangement is clearly visible in the plot.

where $\mathbf{S}_1, \mathbf{S}_2, \ldots, \mathbf{S}_5$ are the covariance matrices of the data in each epoch. We shall then use the first two coordinate values to provide a map of the data showing the relationships between epochs. The necessary R code is:

```
R> skulls_var <- tapply(1:nrow(skulls), skulls$epoch,
+                       function(i) var(skulls[i,-1]))
R> S <- 0
R> for (v in skulls_var) S <- S + 29 * v
R> (S <- S / 149)
```

```
         mb       bh       bl    nh
mb 20.54407  0.03579  0.07696 1.955
bh  0.03579 22.85414  5.06040 2.769
bl  0.07696  5.06040 23.52998 1.103
nh  1.95503  2.76868  1.10291 9.880
```

```
R> skulls_cen <- tapply(1:nrow(skulls), skulls$epoch,
+       function(i) apply(skulls[i,-1], 2, mean))
R> skulls_cen <- matrix(unlist(skulls_cen),
+       nrow = length(skulls_cen), byrow = TRUE)
R> skulls_mah <- apply(skulls_cen, 1,
+       function(cen) mahalanobis(skulls_cen, cen, S))
R> skulls_mah
```

```
        [,1]     [,2]    [,3]    [,4]    [,5]
[1,]  0.00000 0.09355 0.9280  1.9330  2.7712
[2,]  0.09355 0.00000 0.7490  1.6380  2.2357
[3,]  0.92799 0.74905 0.0000  0.4553  0.9360
[4,]  1.93302 1.63799 0.4553  0.0000  0.2253
[5,]  2.77121 2.23571 0.9360  0.2253  0.0000
```

```
R> cmdscale(skulls_mah, k = nrow(skulls_mah) - 1,
+           eig = TRUE)$eig
```

```
[1]  5.113e+00  9.816e-02 -5.551e-16 -1.009e-01 -7.777e-01
```

```
R> skulls_mds <- cmdscale(skulls_mah)
```

The resulting plot is shown in Figure 4.2 and shows that the scaling solution for the skulls data is essentially unidimensional, with this single dimension *time ordering* the five epochs. There appears to be a change in the "shape" of the skulls over time, with maximum breadth increasing and basialiveolar length decreasing.

Our final example of the application of classical multidimensional scaling involves an investigation by Corbet, Cummins, Hedges, and Krzanowski (1970), who report a study of water voles (genus *Arvicola*) in which the aim was to compare British populations of these animals with those in Europe to investigate whether more than one species might be present in Britain. The original data consisted of observations of the presence or absence of 13 characteristics in about 300 water vole skulls arising from six British populations and eight populations from the rest of Europe. Table 4.3 gives a distance matrix derived from the data as described in Corbet et al. (1970).

The following code finds the classical scaling solution and computes the two criteria for assessing the required number of dimensions as described above.

```
R> data("watervoles", package = "HSAUR2")
R> voles_mds <- cmdscale(watervoles, k = 13, eig = TRUE)
R> voles_mds$eig
```

```
 [1]  7.360e-01  2.626e-01  1.493e-01  6.990e-02  2.957e-02
 [6]  1.931e-02  9.714e-17 -1.139e-02 -1.280e-02 -2.850e-02
[11] -4.252e-02 -5.255e-02 -7.406e-02 -1.098e-01
```

Note that some of the eigenvalues are negative. The criterion $P_m^{(1)}$ can be computed by

Table 4.3: watervoles data. Water voles data-dissimilarity matrix.

	Srry	Shrp	Yrks	Prth	Abrd	ElnG	Alps	Ygsl	Grmn	Nrwy	PyrI	PyII	NrtS	SthS
Surrey	0.000													
Shropshire	0.099	0.000												
Yorkshire	0.033	0.022	0.000											
Perthshire	0.183	0.114	0.042	0.000										
Aberdeen	0.148	0.224	0.059	0.068	0.000									
Elean Gamhna	0.198	0.039	0.053	0.085	0.051	0.000								
Alps	0.462	0.266	0.322	0.435	0.268	0.025	0.000							
Yugoslavia	0.628	0.442	0.444	0.406	0.240	0.129	0.014	0.000						
Germany	0.113	0.070	0.046	0.047	0.034	0.002	0.106	0.129	0.000					
Norway	0.173	0.119	0.162	0.331	0.177	0.039	0.089	0.237	0.071	0.000				
Pyrenees I	0.434	0.419	0.339	0.505	0.469	0.390	0.315	0.349	0.151	0.430	0.000			
Pyrenees II	0.762	0.633	0.781	0.700	0.758	0.625	0.469	0.618	0.440	0.538	0.607	0.000		
North Spain	0.530	0.389	0.482	0.579	0.597	0.498	0.374	0.562	0.247	0.383	0.387	0.084	0.000	
South Spain	0.586	0.435	0.550	0.530	0.552	0.509	0.369	0.471	0.234	0.346	0.456	0.090	0.038	0.000

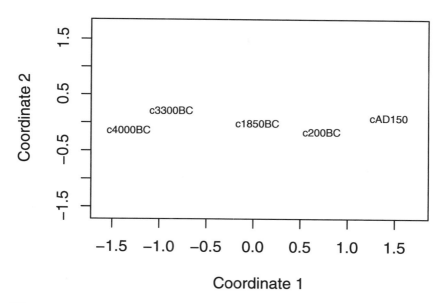

Fig. 4.2. Two-dimensional solution from classical MDS applied to Mahalanobis distances between epochs for the skull data.

```
R> cumsum(abs(voles_mds$eig))/sum(abs(voles_mds$eig))
```

```
[1] 0.4605 0.6248 0.7182 0.7619 0.7804 0.7925 0.7925 0.7996
[9] 0.8077 0.8255 0.8521 0.8850 0.9313 1.0000
```

and the criterion $P_m^{(2)}$ is

```
R> cumsum((voles_mds$eig)^2)/sum((voles_mds$eig)^2)
```

```
[1] 0.8179 0.9220 0.9557 0.9631 0.9644 0.9649 0.9649 0.9651
[9] 0.9654 0.9666 0.9693 0.9735 0.9818 1.0000
```

Here the two criteria for judging the number of dimensions necessary to give an adequate fit to the data are quite different. The second criterion would suggest that two dimensions is adequate, but use of the first would suggest perhaps that three or even four dimensions might be required. Here we shall be guided by the second fit index and the two-dimensional solution that can be plotted by extracting the coordinates from the **points** element of the **voles_mds** object; the plot is shown in Figure 4.3.

It appears that the six British populations are close to populations living in the Alps, Yugoslavia, Germany, Norway, and Pyrenees I (consisting

```
R> x <- voles_mds$points[,1]
R> y <- voles_mds$points[,2]
R> plot(x, y, xlab = "Coordinate 1", ylab = "Coordinate 2",
+       xlim = range(x)*1.2, type = "n")
R> text(x, y, labels = colnames(watervoles), cex = 0.7)
```

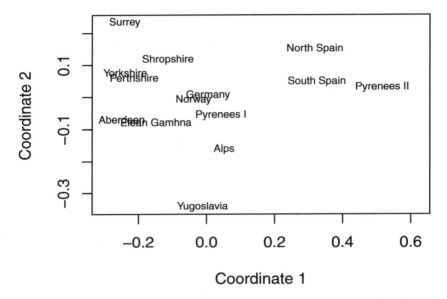

Fig. 4.3. Two-dimensional solution from classical multidimensional scaling of the distance matrix for water vole populations.

of the species *Arvicola terrestris*) but rather distant from the populations in Pyrenees II, North Spain and South Spain (species *Arvicola sapidus*). This result would seem to imply that *Arvicola terrestris* might be present in Britain but it is less likely that this is so for *Arvicola sapidus*. But here, as the two-dimensional fit may not be entirely what is needed to represent the observed distances, we shall investigate the solution in a little more detail using the *minimum spanning tree*.

The *minimum spanning tree* is defined as follows. Suppose n points are given (possibly in many dimensions). Then a tree spanning these points (i.e., a spanning tree) is any set of straight line segments joining pairs of points such that

- o closed loops occur,
- every point is visited at least one time, and

- the tree is connected (i.e., it has paths between any pairs of points).

The length of the tree is defined to be the sum of the length of its segments, and when a set of n points and the lengths of all $\binom{n}{2}$ segments are given, then the minimum spanning tree is defined as the spanning tree with minimum length. Algorithms to find the minimum spanning tree of a set of n points given the distances between them are given in Prim (1957) and Gower and Ross (1969).

The links of the minimum spanning tree (of the spanning tree) of the proximity matrix of interest may be plotted onto the two-dimensional scaling representation in order to identify possible distortions produced by the scaling solutions. Such distortions are indicated when nearby points on the plot are not linked by an edge of the tree.

To find the minimum spanning tree of the water vole proximity matrix, the function mst from the package **ape** (Paradis, Bolker, Claude, Cuong, Desper, Durand, Dutheil, Gascuel, Heibl, Lawson, Lefort, Legendre, Lemon, Noel, Nylander, Opgen-Rhein, Schliep, Strimmer, and de Vienne 2010) can be used, and we can plot the minimum spanning tree on the two-dimensional scaling solution as shown in Figure 4.4.

The plot indicates, for example, that the apparent closeness of the populations in Germany and Norway, suggested by the points representing them in the MDS solution, does not accurately reflect their calculated dissimilarity; the links of the minimum spanning tree show that the Aberdeen and Elean Gamhna populations are actually more similar to the German water voles than those from Norway. This suggests that the two-dimensional solution may not give an adequate representation of the whole distance matrix.

4.5 Non-metric multidimensional scaling

In some psychological work and in market research, proximity matrices arise from asking human subjects to make judgements about the similarity or dissimilarity of objects or stimuli of interest. When collecting such data, the investigator may feel that realistically subjects are only able to give "ordinal" judgements; for example, when comparing a range of colours they might be able to specify with some confidence that one colour is brighter than another but would be far less confident if asked to put a value to how much brighter. Such considerations led, in the 1960s, to the search for a method of multidimensional scaling that uses only the *rank order* of the proximities to produce a spatial representation of them. In other words, a method was sought that would be invariant under *monotonic transformations* of the observed proximity matrix; i.e., the derived coordinates will remain the same if the numerical values of the observed proximities are changed but their rank order is not. Such a method was proposed in landmark papers by Shepard (1962a,b) and by Kruskal (1964a). The quintessential component of the method proposed in

```
R> library("ape")
R> st <- mst(watervoles)
R> plot(x, y, xlab = "Coordinate 1", ylab = "Coordinate 2",
+        xlim = range(x)*1.2, type = "n")
R> for (i in 1:nrow(watervoles)) {
+        w1 <- which(st[i, ] == 1)
+        segments(x[i], y[i], x[w1], y[w1])
+ }
R> text(x, y, labels = colnames(watervoles), cex = 0.7)
```

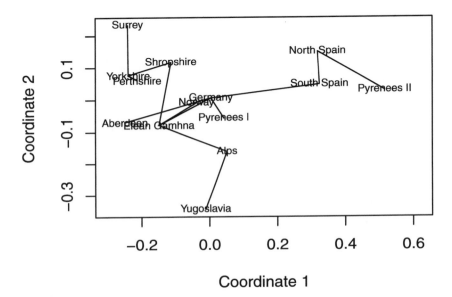

Fig. 4.4. Minimum spanning tree for the `watervoles` data plotted onto the classical scaling two-dimensional solution.

these papers is that the coordinates in the spatial representation of the observed proximities give rise to fitted distances, d_{ij}, and that these distances are related to a set of numbers which we will call *disparities*, $\hat{d}_{ij}$, by the formula $d_{ij} = \hat{d}_{ij} + \epsilon_{ij}$, where the ϵ_{ij} are error terms representing errors of measurement plus distortion errors arising because the distances do not correspond to a configuration in the particular number of dimensions chosen. The disparities are monotonic with the observed proximities and, subject to this constraint, resemble the fitted distances as closely as possible. In general, only a weak monotonicity constraint is applied, so that if, say, the observed dissimilarities, δ_{ij} are ranked from lowest to highest to give

$$\delta_{i_1 j_1} < \delta_{i_2 j_2} < \cdots < \delta_{i_N j_N},$$

where $N = n(n-1)/2$, then

$$\hat{d}_{i_1 j_1} \leq \hat{d}_{i_2 j_2} \leq \cdots \leq \hat{d}_{i_N j_N}.$$

Monotonic regression (see Barlow, Bartholomew, Bremner, and Brunk 1972) is used to find the disparities, and then the required coordinates in the spatial representation of the observed dissimilarities, which we denote by $\hat{\mathbf{X}}(n \times m)$, are found by minimising a criterion, S, known as *Stress*, which is a function of $\hat{\mathbf{X}}(n \times m)$ and is defined as

$$S(\hat{\mathbf{X}}) = \min \sum_{i<j} (\hat{d}_{ij} - d_{ij})^2 / \sum_{i<j} d_{ij}^2,$$

where the minimum is taken over $\hat{d}_{ij}$ such that $\hat{d}_{ij}$ is monotonic with the observed dissimilarities. In essence, Stress represents the extent to which the rank order of the fitted distances disagrees with the rank order of the observed dissimilarities. The denominator in the formula for Stress is chosen to make the final spatial representation invariant under changes of scale; i.e., uniform stretching or shrinking. An algorithm to minimise Stress and so find the coordinates of the required spatial representation is described in a second paper by Kruskal (1964b). For each value of the number of dimensions, m, in the spatial configuration, the configuration that has the smallest Stress is called the best-fitting configuration in m dimensions, S_m, and a rule of thumb for judging the fit as given by Kruskal is $S_m \geq 20\%$, poor, $S_m = 10\%$, fair, $S_m \leq 5\%$, good; and $S_m = 0$, perfect (this only occurs if the rank order of the fitted distances matches the rank order of the observed dissimilarities and event, which is, of course, very rare in practise). We will now look at some applications of non-metric multidimensional scaling.

4.5.1 House of Representatives voting

Romesburg (1984) gives a set of data that shows the number of times 15 congressmen from New Jersey voted differently in the House of Representatives on 19 environmental bills. Abstentions are not recorded, but two congressmen abstained more frequently than the others, these being Sandman (nine abstentions) and Thompson (six abstentions). The data are available in Table 4.4 and of interest is the question of whether party affiliations can be detected in the data.

We shall now apply non-metric scaling to the voting behaviour shown in Table 4.4. Non-metric scaling is available with the function isoMDS from the package **MASS** (Venables and Ripley 2010, 2002)

```
R> library("MASS")
R> data("voting", package = "HSAUR2")
R> voting_mds <- isoMDS(voting)
```

and we again plot the two-dimensional solution (Figure 4.5). The figure suggests that voting behaviour is essentially along party lines, although there is more variation among Republicans. The voting behaviour of one of the Republicans (Rinaldo) seems to be closer to his Democratic colleagues rather than to the voting behaviour of other Republicans.

Table 4.4: **voting** data. House of Representatives voting data; (R) is short for Republican, (D) for Democrat.

	Hnt	Snd	Hwr	Thm	Fry	Frs	Wdn	Roe	Hlt	Rdn	Mns	Rnl	Mrz	Dnl	Ptt
Hunt(R)	0														
Sandman(R)	8	0													
Howard(D)	15	17	0												
Thompson(D)	15	12	9	0											
Freylinghuysen(R)	10	13	16	14	0										
Forsythe(R)	9	13	12	12	8	0									
Widnall(R)	7	12	15	13	9	7	0								
Roe(D)	15	16	5	10	13	12	17	0							
Heltoski(D)	16	17	5	8	14	11	16	4	0						
Rodino(D)	14	15	6	8	12	10	15	5	3	0					
Minish(D)	15	16	5	8	12	9	14	5	2	1	0				
Rinaldo(R)	16	17	4	6	12	10	15	3	1	2	1	0			
Maraziti(R)	7	13	11	15	10	6	10	12	13	11	12	12	0		
Daniels(D)	11	12	10	10	11	6	11	7	7	4	5	6	9	0	
Patten(D)	13	16	7	7	11	10	13	6	5	6	5	4	13	9	0

The quality of a multidimensional scaling can be assessed informally by plotting the original dissimilarities and the distances obtained from a multidimensional scaling in a scatterplot, a so-called Shepard diagram. For the **voting** data, such a plot is shown in Figure 4.6. In an ideal situation, the points fall on the bisecting line; in our case, some deviations are observable.

4.5.2 Judgements of World War II leaders

As the first example of the application of non-metric multidimensional scaling, we shall use the method to get a spatial representation of the judgements of the dissimilarities in ideology of a number of world leaders and politicians prominent at the time of the Second World War, shown in Table 4.5. The subject made judgements on a nine-point scale, with the anchor points of the scale, 1 and 9, being described as indicating "very similar" and "very dissimilar", respectively; this was all the subject was told about the scale.

```
R> x <- voting_mds$points[,1]
R> y <- voting_mds$points[,2]
R> plot(x, y, xlab = "Coordinate 1", ylab = "Coordinate 2",
+       xlim = range(voting_mds$points[,1])*1.2, type = "n")
R> text(x, y, labels = colnames(voting), cex = 0.6)
R> voting_sh <- Shepard(voting[lower.tri(voting)],
+                       voting_mds$points)
```

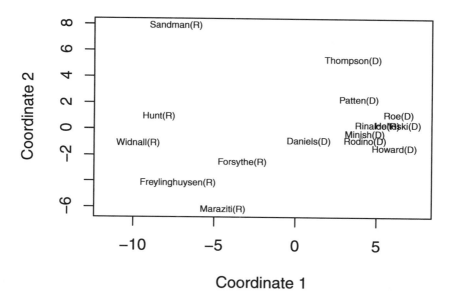

Fig. 4.5. Two-dimensional solution from non-metric multidimensional scaling of distance matrix for voting matrix.

Table 4.5: `WWIIleaders` data. Subjective distances between WWII leaders.

	Htl	Mss	Chr	Esn	Stl	Att	Frn	DGl	MT-	Trm	Chm	Tit
Hitler	0											
Mussolini	3	0										
Churchill	4	6	0									
Eisenhower	7	8	4	0								
Stalin	3	5	6	8	0							
Attlee	8	9	3	9	8	0						
Franco	3	2	5	7	6	7	0					
De Gaulle	4	4	3	5	6	5	4	0				

Table 4.5: WWIIleaders data (continued).

	Htl	Mss	Chr	Esn	Stl	Att	Frn	DGl	MT-	Trm	Chm	Tit
Mao Tse-Tung	8	9	8	9	6	9	8	7	0			
Truman	9	9	5	4	7	8	8	4	4	0		
Chamberlin	4	5	5	4	7	2	2	5	9	5	0	
Tito	7	8	2	4	7	8	3	2	4	5	7	0

The non-metric multidimensional scaling applied to these distances is

```
R> (WWII_mds <- isoMDS(WWIIleaders))
```

```
initial  value 20.504211
iter   5 value 15.216103
iter   5 value 15.207237
iter   5 value 15.207237
final  value 15.207237
converged
$points
                [,1]     [,2]
Hitler       -2.5820 -1.75961
Mussolini    -3.8807 -1.24756
Churchill     0.3110  1.46671
Eisenhower    2.9852  2.87822
Stalin       -1.4274 -3.75699
Attlee       -2.1067  5.07317
Franco       -2.8590  0.07878
De Gaulle     0.6591 -0.20656
Mao Tse-Tung  4.1605 -4.57583
Truman        4.4962  0.29294
Chamberlin   -2.1420  2.75877
Tito          2.3859 -1.00203
```

```
$stress
[1] 15.21
```

The two-dimensional solution appears in Figure 4.7. Clearly, the three fascists group together as do the three British prime ministers. Stalin and Mao Tse-Tung are more isolated compared with the other leaders. Eisenhower seems more related to the British government than to his own President Truman. Interestingly, de Gaulle is placed in the center of the MDS solution.

```
R> plot(voting_sh, pch = ".", xlab = "Dissimilarity",
+       ylab = "Distance", xlim = range(voting_sh$x),
+       ylim = range(voting_sh$x))
R> lines(voting_sh$x, voting_sh$yf, type = "S")
```

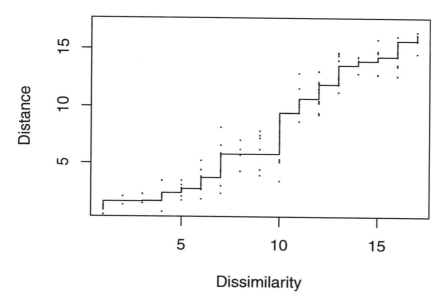

Fig. 4.6. The Shepard diagram for the `voting` data shows some discrepancies between the original dissimilarities and the multidimensional scaling solution.

4.6 Correspondence analysis

A form of multidimensional scaling known as *correspondence analysis*, which is essentially an approach to constructing a spatial model that displays the associations among a set of categorical variables, will be the subject of this section. Correspondence analysis has a relatively long history (see de Leeuw 1983) but for a long period was only routinely used in France, largely due to the almost evangelical efforts of Benzécri (1992). But nowadays the method is used rather more widely and is often applied to supplement, say, a standard chi-squared test of independence for two categorical variables forming a contingency table.

Mathematically, correspondence analysis can be regarded as either

- a method for decomposing the chi-squared statistic used to test for independence in a contingency table into components corresponding to different dimensions of the heterogeneity between its columns, or

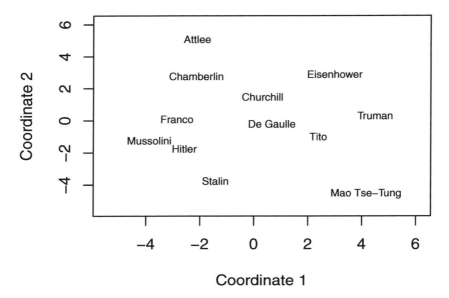

Fig. 4.7. Non-metric multidimensional scaling of perceived distances of World War II leaders.

- a method for simultaneously assigning a scale to rows and a separate scale to columns so as to maximise the correlation between the two scales.

Quintessentially, however, correspondence analysis is a technique for displaying multivariate (most often bivariate) categorical data graphically by deriving coordinates to represent the categories of both the row and column variables, which may then be plotted so as to display the pattern of association between the variables graphically. A detailed account of correspondence analysis is given in Greenacre (2007), where its similarity to principal components and the biplot is stressed. Here we give only accounts of the method demonstrating the use of classical multidimensional scaling to get a two-dimensional map to represent a set of data in the form of a two-dimensional contingency table.

The general two-dimensional contingency table in which there are r rows and c columns can be written as

	y		
	1 $\;\ldots\;$ c		
1	$n_{11}\;\ldots\;n_{1c}$	$n_{1\cdot}$
2	$n_{21}\;\ldots\;n_{2c}$	$n_{2\cdot}$
$x\;\vdots$	$\vdots\;\ldots\;\vdots$	$\vdots$
r	$n_{r1}\;\ldots\;n_{rc}$	$n_{r\cdot}$
	$n_{\cdot1}\;\ldots\;n_{\cdot c}$	n

using an obvious dot notation for summing the counts in the contingency table over rows or over columns. From this table we can construct tables of column proportions and row proportions given by

Column proportions $p_{ij}^{c} = n_{ij}/n_{\cdot j}$,
Row proportions $p_{ij}^{r} = n_{ij}/n_{i\cdot}$.

What is known as the *chi-squared distance* between columns i and j is defined as

$$d_{ij}^{(\text{cols})} = \sum_{k=1}^{r} \frac{1}{p_{k\cdot}} (p_{ki}^{c} - p_{kj}^{c})^2,$$

where

$$p_{k\cdot} = n_{k\cdot}/n.$$

The chi-square distance is seen to be a weighted Euclidean distance based on column proportions. It will be zero if the two columns have the same values for these proportions. It can also be seen from the weighting factors, $1/p_{k\cdot}$, that rare categories of the column variable have a greater influence on the distance than common ones.

A similar distance measure can be defined for rows i and j as

$$d_{ij}^{(\text{rows})} = \sum_{k=1}^{c} \frac{1}{p_{\cdot k}} (p_{ik}^{r} - p_{jk}^{r})^2,$$

where

$$p_{\cdot k} = n_{\cdot k}/n.$$

A correspondence analysis "map" of the data can be found by applying classical MDS to each distance matrix in turn and plotting usually the first two coordinates for column categories and those for row categories on the same diagram, suitably labelled to differentiate the points representing row categories from those representing column categories. The resulting diagram is interpreted by examining the positions of the points representing the row categories and the column categories. The relative values of the coordinates of these points reflect associations between the categories of the row variable and the categories of the column variable. Assuming that a two-dimensional solution provides an adequate fit for the data (see Greenacre 1992), row points that are close together represent row categories that have similar profiles (conditional distributions) across columns. Column points that are close together indicate columns with similar profiles (conditional distributions) down the rows. Finally, row points that lie close to column points represent a row/column combination that occurs more frequently in the table than would be expected if the row and column variables were independent. Conversely, row and column points that are distant from one another indicate a cell in the table where the count is lower than would be expected under independence.

We will now look at a single simple example of the application of correspondence analysis.

4.6.1 Teenage relationships

Consider the data shown in Table 4.6 concerned with the influence of a girl's age on her relationship with her boyfriend. In this table, each of 139 girls has been classified into one of three groups:

- no boyfriend;
- boyfriend/no sexual intercourse; or
- boyfriend/sexual intercourse.

In addition, the age of each girl was recorded and used to divide the girls into five age groups.

Table 4.6: `teensex` data. The influence of age on relationships with boyfriends.

Boyfriend	Age				
	<16	16-17	17-18	18-19	19-20
No boyfriend	21	21	14	13	8
Boyfriend no sex	8	9	6	8	2
Boyfriend sex	2	3	4	10	10

The calculation of the two-dimensional classical multidimensional scaling solution based on the row- and column-wise chi-squared distance measure can be computed via `cmdscale()`; however, we first have to compute the necessary row and column distance matrices, and we will do this by setting up a small convenience function as follows:

```
R> D <- function(x) {
+       a <- t(t(x) / colSums(x))
+       ret <- sqrt(colSums((a[,rep(1:ncol(x), ncol(x))] -
+           a[, rep(1:ncol(x), rep(ncol(x), ncol(x)))])^2 *
+               sum(x) / rowSums(x)))
+       matrix(ret, ncol = ncol(x))
+ }
R> (dcols <- D(teensex))
```

```
         [,1]     [,2]    [,3]    [,4]    [,5]
[1,]  0.00000  0.08537  0.2574  0.6629  1.0739
[2,]  0.08537  0.00000  0.1864  0.5858  1.0142
[3,]  0.25743  0.18644  0.0000  0.4066  0.8295
[4,]  0.66289  0.58581  0.4066  0.0000  0.5067
[5,]  1.07385  1.01423  0.8295  0.5067  0.0000
```

R> (drows <- D(t(teensex)))

```
         [,1]     [,2]    [,3]
[1,]  0.0000  0.2035  0.9275
[2,]  0.2035  0.0000  0.9355
[3,]  0.9275  0.9355  0.0000
```

Applying classical MDS to each of these distance matrices gives the required two-dimensional coordinates with which to construct our "map" of the data. Plotting those with suitable labels and with the axes suitably scaled to reflect the greater variation on dimension one than on dimension two (see Greenacre 1992) is achieved using the R code presented with Figure 4.8.

The points representing the age groups in Figure 4.8 give a two-dimensional representation in which the Euclidean distance between two points represents the chi-squared distance between the corresponding age groups (and similarly for the points representing the type of relationship). For a contingency table with r rows and c columns, it can be shown that the chi-squared distances can be represented *exactly* in $\min r - 1, c - 1$ dimensions; here, since $r = 3$ and $c = 5$, this means that the Euclidean distances in Figure 4.8 will actually *equal* the corresponding chi-squared distances (readers might like to check that this is the case as an exercise). When both r and c are greater than three, an exact two-dimensional representation of the chi-squared distances is not possible. In such cases, the derived two-dimensional coordinates will give only an approximate representation, and so the question of the adequacy of the fit will need to be addressed. In some of these cases, more than two dimensions may be required to give an acceptable fit (again see Greenacre 1992, for details).

Examining the plot in Figure 4.8, we see that it tells the age-old story of girls travelling through their teenage years, initially having no boyfriend, then acquiring a boyfriend, and then having sex with their boyfriend, a story that has broken the hearts of fathers everywhere, at least temporarily, until their wives suggest they reflect back to the time when they themselves were teenagers.

4.7 Summary

Multidimensional scaling and correspondence analysis both aim to help in understanding particular types of data by displaying the data graphically.

```
R> r1 <- cmdscale(dcols, eig = TRUE)
R> c1 <- cmdscale(drows, eig = TRUE)
R> plot(r1$points, xlim = range(r1$points[,1], c1$points[,1]) * 1.5,
+       ylim = range(r1$points[,1], c1$points[,1]) * 1.5, type = "n",
+       xlab = "Coordinate 1", ylab = "Coordinate 2", lwd = 2)
R> text(r1$points, labels = colnames(teensex), cex = 0.7)
R> text(c1$points, labels = rownames(teensex), cex = 0.7)
R> abline(h = 0, lty = 2)
R> abline(v = 0, lty = 2)
```

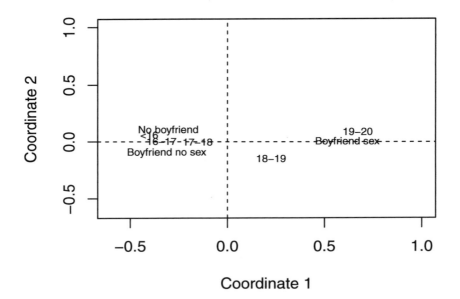

Fig. 4.8. Correspondence analysis for teenage relationship data.

Multidimensional scaling applied to proximity matrices is often useful in uncovering the dimensions on which similarity judgements are made, and correspondence analysis often allows more insight into the pattern of relationships in a contingency table than a simple chi-squared test.

4.8 Exercises

Ex. 4.1 Consider 51 objects $O_1, \ldots, O_{51}$ assumed to be arranged along a straight line with the jth object being located at a point with coordinate j. Define the similarity s_{ij} between object i and object j as

$$s_{ij} = \begin{cases} 9 & \text{if } i = j \\ 8 & \text{if } 1 \leq |i - j| \leq 3 \\ 7 & \text{if } 4 \leq |i - j| \leq 6 \\ \quad \cdots \\ 1 & \text{if } 22 \leq |i - j| \leq 24 \\ 0 & \text{if } |i - j| \geq 25. \end{cases}$$

Convert these similarities into dissimilarities (δ_{ij}) by using

$$\delta_{ij} = \sqrt{s_{ii} + s_{jj} - 2s_{ij}}$$

and then apply classical multidimensional scaling to the resulting dissimilarity matrix. Explain the shape of the derived two-dimensional solution.

Ex. 4.2 Write an R function to calculate the chi-squared distance matrices for both rows and columns in a two-dimensional contingency table.

Ex. 4.3 In Table 4.7 (from Kaufman and Rousseeuw 1990), the dissimilarity matrix of 18 species of garden flowers is shown. Use some form of multidimensional scaling to investigate which species share common properties.

Table 4.7: **gardenflowers** data. Dissimilarity matrix of 18 species of garden-flowers.

	Bgn	Brm	Cml	Dhl	F-	Fch	Grn	Gld	Hth	Hyd	Irs	Lly	L-	Pny	Pnc	Rdr	Scr	Tlp
Begonia	0.00																	
Broom	0.91	0.00																
Camellia	0.49	0.67	0.00															
Dahlia	0.47	0.59	0.59	0.00														
Forget-me-not	0.43	0.90	0.57	0.61	0.00													
Fuchsia	0.23	0.79	0.29	0.52	0.44	0.00												
Geranium	0.31	0.70	0.54	0.44	0.54	0.24	0.00											
Gladiolus	0.49	0.57	0.71	0.26	0.49	0.68	0.49	0.00										
Heather	0.57	0.57	0.57	0.89	0.50	0.61	0.70	0.77	0.00									
Hydrangea	0.76	0.58	0.58	0.62	0.39	0.61	0.86	0.70	0.55	0.00								
Iris	0.32	0.77	0.63	0.75	0.46	0.52	0.60	0.63	0.46	0.47	0.00							
Lily	0.51	0.69	0.69	0.53	0.51	0.65	0.77	0.47	0.51	0.39	0.36	0.00						
Lily-of-the-valley	0.59	0.75	0.75	0.77	0.35	0.63	0.72	0.65	0.35	0.41	0.45	0.24	0.00					
Peony	0.37	0.68	0.68	0.38	0.52	0.48	0.63	0.49	0.52	0.39	0.37	0.17	0.39	0.00				
Pink carnation	0.74	0.54	0.70	0.58	0.54	0.74	0.50	0.49	0.36	0.52	0.60	0.48	0.39	0.49	0.00			
Red rose	0.84	0.41	0.75	0.37	0.82	0.71	0.61	0.64	0.81	0.43	0.84	0.62	0.67	0.47	0.45	0.00		
Scotch rose	0.94	0.20	0.70	0.48	0.77	0.83	0.74	0.45	0.77	0.38	0.80	0.58	0.62	0.57	0.40	0.21	0.00	
Tulip	0.44	0.50	0.79	0.48	0.59	0.68	0.47	0.22	0.59	0.92	0.59	0.67	0.72	0.67	0.61	0.85	0.67	0.00

5

Exploratory Factor Analysis

5.1 Introduction

In many areas of psychology, and other disciplines in the behavioural sciences, often it is not possible to measure directly the concepts of primary interest. Two obvious examples are *intelligence* and *social class*. In such cases, the researcher is forced to examine the concepts *indirectly* by collecting information on variables that *can* be measured or observed directly and can also realistically be assumed to be indicators, in some sense, of the concepts of real interest. The psychologist who is interested in an individual's "intelligence", for example, may record examination scores in a variety of different subjects in the expectation that these scores are dependent in some way on what is widely regarded as "intelligence" but are also subject to random errors. And a sociologist, say, concerned with people's "social class" might pose questions about a person's occupation, educational background, home ownership, etc., on the assumption that these do reflect the concept he or she is really interested in.

Both "intelligence" and "social class" are what are generally referred to as *latent variables*–i.e., concepts that cannot be measured directly but can be assumed to relate to a number of measurable or *manifest variables*. The method of analysis most generally used to help uncover the relationships between the assumed latent variables and the manifest variables is *factor analysis*. The model on which the method is based is essentially that of multiple regression, except now the manifest variables are regressed on the unobservable latent variables (often referred to in this context as *common factors*), so that direct estimation of the corresponding regression coefficients (*factor loadings*) is not possible.

A point to be made at the outset is that factor analysis comes in two distinct varieties. The first is *exploratory factor analysis*, which is used to investigate the relationship between manifest variables and factors without making any assumptions about which manifest variables are related to which factors. The second is *confirmatory factor analysis* which is used to test whether a specific factor model postulated a priori provides an adequate fit for the covariances or correlations between the manifest variables. In this chapter, we shall consider only exploratory factor analysis. Confirmatory factor analysis will be the subject of Chapter 7.

Exploratory factor analysis is often said to have been introduced by Spearman (1904), but this is only partially true because Spearman proposed only the one-factor model as described in the next section. Fascinating accounts of the history of factor analysis are given in Thorndike (2005) and Bartholomew (2005).

5.2 A simple example of a factor analysis model

To set the scene for the k-factor analysis model to be described in the next section, we shall in this section look at a very simple example in which there is only a single factor.

Spearman considered a sample of children's examination marks in three subjects, Classics (x_1), French (x_2), and English (x_3), from which he calculated the following correlation matrix for a sample of children:

$$\mathbf{R} = \begin{matrix} \text{Classics} \\ \text{French} \\ \text{English} \end{matrix} \begin{pmatrix} 1.00 & & \\ 0.83 & 1.00 & \\ 0.78 & 0.67 & 1.00 \end{pmatrix}.$$

If we assume a single factor, then the *single-factor model* is specified as follows:

$$x_1 = \lambda_1 f + u_1,$$
$$x_2 = \lambda_2 f + u_2,$$
$$x_3 = \lambda_3 f + u_3.$$

We see that the model essentially involves the simple linear regression of each observed variable on the single common factor. In this example, the underlying latent variable or common factor, f, might possibly be equated with intelligence or general intellectual ability. The terms λ_1, λ_2, and λ_3 which are essentially regression coefficients are, in this context, known as *factor loadings*, and the terms u_1, u_2, and u_3 represent random disturbance terms and will have small variances if their associated observed variable is closely related to the underlying latent variable. The variation in u_i actually consists of two parts, the extent to which an individual's ability at Classics, say, differs from his or her general ability and the extent to which the examination in Classics is only an approximate measure of his or her ability in the subject. In practise no attempt is made to disentangle these two parts.

We shall return to this simple example later when we consider how to estimate the parameters in the factor analysis model. Before this, however, we need to describe the factor analysis model itself in more detail. The description follows in the next section.

5.3 The k-factor analysis model

The basis of factor analysis is a regression model linking the manifest variables to a set of unobserved (and unobservable) latent variables. In essence the model assumes that the observed relationships between the manifest variables (as measured by their covariances or correlations) are a result of the relationships of these variables to the latent variables. (Since it is the covariances or correlations of the manifest variables that are central to factor analysis, we can, in the description of the mathematics of the method given below, assume that the manifest variables all have zero mean.)

To begin, we assume that we have a set of observed or manifest variables, $\mathbf{x}^\top = (x_1, x_2, \ldots, x_q)$, assumed to be linked to k unobserved latent variables or common factors $f_1, f_2, \ldots, f_k$, where $k < q$, by a regression model of the form

$$x_1 = \lambda_{11} f_1 + \lambda_{12} f_2 + \cdots + \lambda_{1k} f_k + u_1,$$
$$x_2 = \lambda_{21} f_1 + \lambda_{22} f_2 + \cdots + \lambda_{2k} f_k + u_2,$$
$$\vdots$$
$$x_q = \lambda_{q1} f_1 + \lambda_{q2} f_2 + \cdots + \lambda_{qk} f_k + u_q.$$

The λ_js are essentially the regression coefficients of the x-variables on the common factors, but in the context of factor analysis these regression coefficients are known as the factor loadings and show how each observed variable, x_i, depends on the common factors. The factor loadings are used in the interpretation of the factors; i.e., larger values relate a factor to the corresponding observed variables and from these we can often, but not always, infer a meaningful description of each factor (we will give examples later).

The regression equations above may be written more concisely as

$$\mathbf{x} = \Lambda \mathbf{f} + \mathbf{u},$$

where

$$\Lambda = \begin{pmatrix} \lambda_{11} & \cdots & \lambda_{1k} \\ \vdots & & \vdots \\ \lambda_{q1} & \cdots & \lambda_{qk} \end{pmatrix}, \quad \mathbf{f} = \begin{pmatrix} f_1 \\ \vdots \\ f_q \end{pmatrix}, \quad \mathbf{u} = \begin{pmatrix} u_1 \\ \vdots \\ u_q \end{pmatrix}.$$

We assume that the random disturbance terms $u_1, \ldots, u_q$ are uncorrelated with each other and with the factors $f_1, \ldots, f_k$. (The elements of $\mathbf{u}$ are specific to each x_i and hence are generally better known in this context as *specific variates*.) The two assumptions imply that, given the values of the common factors, the manifest variables are independent; that is, the correlations of the observed variables arise from their relationships with the common factors. Because the factors are unobserved, we can fix their locations and scales arbitrarily and we shall assume they occur in standardised form with mean zero and standard deviation one. We will also assume, initially at least, that the factors are uncorrelated with one another, in which case the factor loadings are the *correlations* of the manifest variables and the factors. With these additional assumptions about the factors, the factor analysis model implies that the variance of variable x_i, σ_i^2, is given by

$$\sigma_i^2 = \sum_{j=1}^{k} \lambda_{ij}^2 + \psi_i,$$

where ψ_i is the variance of u_i. Consequently, we see that the factor analysis model implies that the variance of each observed variable can be split into two parts: the first, h_i^2, given by $h_i^2 = \sum_{j=1}^{k} \lambda_{ij}^2$, is known as the *communality* of the variable and represents the variance shared with the other variables via the common factors. The second part, ψ_i, is called the *specific* or *unique* variance and relates to the variability in x_i not shared with other variables. In addition, the factor model leads to the following expression for the covariance of variables x_i and x_j:

$$\sigma_{ij} = \sum_{l=1}^{k} \lambda_{il} \lambda_{jl}.$$

We see that the covariances are *not* dependent on the specific variates in any way; it is the common factors only that aim to account for the relationships between the manifest variables.

The results above show that the k-factor analysis model implies that the population covariance matrix, $\boldsymbol{\Sigma}$, of the observed variables has the form

$$\boldsymbol{\Sigma} = \boldsymbol{\Lambda}\boldsymbol{\Lambda}^{\top} + \boldsymbol{\Psi},$$

where

$$\boldsymbol{\Psi} = \mathrm{diag}(\boldsymbol{\Psi}_i).$$

The converse also holds: if $\boldsymbol{\Sigma}$ can be decomposed into the form given above, then the k-factor model holds for **x**. In practise, $\boldsymbol{\Sigma}$ will be estimated by the sample covariance matrix **S** and we will need to obtain estimates of $\boldsymbol{\Lambda}$ and $\boldsymbol{\Psi}$ so that the observed covariance matrix takes the form required by the model (see later in the chapter for an account of estimation methods). We will also need to determine the value of k, the number of factors, so that the model provides an adequate fit for **S**.

5.4 Scale invariance of the k-factor model

Before describing both estimation for the k-factor analysis model and how to determine the appropriate value of k, we will consider how rescaling the x variables affects the factor analysis model. Rescaling the x variables is equivalent to letting $\mathbf{y} = \mathbf{Cx}$, where $\mathbf{C} = \mathrm{diag}(c_i)$ and the c_i, $i = 1, \ldots, q$ are the scaling values. If the k-factor model holds for x with $\boldsymbol{\Lambda} = \boldsymbol{\Lambda}_x$ and $\boldsymbol{\Psi} = \boldsymbol{\Psi}_x$, then

$$\mathbf{y} = \mathbf{C}\boldsymbol{\Psi}_x\mathbf{f} + \mathbf{C}\mathbf{u}$$

and the covariance matrix of $\mathbf{y}$ implied by the factor analysis model for $\mathbf{x}$ is

$$\mathsf{Var}(\mathbf{y}) = \mathbf{C}\boldsymbol{\Sigma}\mathbf{C} = \mathbf{C}\boldsymbol{\Lambda}_x\mathbf{C} + \mathbf{C}\boldsymbol{\Psi}_x\mathbf{C}.$$

So we see that the k-factor model also holds for $\mathbf{y}$ with factor loading matrix $\boldsymbol{\Lambda}_y = \mathbf{C}\boldsymbol{\Lambda}_x$ and specific variances $\boldsymbol{\Psi}_y = \mathbf{C}\boldsymbol{\Psi}_x\mathbf{C} = c_i^2\psi_i$. So the factor loading matrix for the scaled variables $\mathbf{y}$ is found by scaling the factor loading matrix of the original variables by multiplying the ith row of $\boldsymbol{\Lambda}_x$ by c_i and similarly for the specific variances. Thus factor analysis is essentially unaffected by the rescaling of the variables. In particular, if the rescaling factors are such that $c_i = 1/s_i$, where s_i is the standard deviation of the x_i, then the rescaling is equivalent to applying the factor analysis model to the correlation matrix of the x variables and the factor loadings and specific variances that result can be found simply by scaling the corresponding loadings and variances obtained from the covariance matrix. Consequently, the factor analysis model can be applied to either the covariance matrix or the correlation matrix because the results are essentially equivalent. (Note that this is *not* the same as when using principal components analysis, as pointed out in Chapter 3, and we will return to this point later in the chapter.)

5.5 Estimating the parameters in the k-factor analysis model

To apply the factor analysis model outlined in the previous section to a sample of multivariate observations, we need to estimate the parameters of the model in some way. These parameters are the factor loadings and specific variances, and so the estimation problem in factor analysis is essentially that of finding $\hat{\boldsymbol{\Lambda}}$ (the estimated factor loading matrix) and $\hat{\boldsymbol{\Psi}}$ (the diagonal matrix containing the estimated specific variances), which, assuming the factor model outlined in Section 5.3, reproduce as accurately as possible the sample covariance matrix, $\boldsymbol{S}$. This implies

$$\mathbf{S} \approx \hat{\boldsymbol{\Lambda}}\hat{\boldsymbol{\Lambda}}^\top + \hat{\boldsymbol{\Psi}}.$$

Given an estimate of the factor loading matrix, $\hat{\boldsymbol{\Lambda}}$, it is clearly sensible to estimate the specific variances as

$$\hat{\psi}_i = s_i^2 - \sum_{j=1}^{k} \hat{\lambda}_{ij}^2, \quad i = 1, \ldots, q$$

so that the diagonal terms in $\mathbf{S}$ are estimated exactly.

Before looking at methods of estimation used in practise, we shall for the moment return to the simple single-factor model considered in Section 5.2 because in this case estimation of the factor loadings and specific variances is very simple, the reason being that in this case the number of parameters in the model, 6 (three factor loadings and three specific variances), is equal to the number of independent elements in $\mathbf{R}$ (the three correlations and the three diagonal standardised variances), and so by equating elements of the observed correlation matrix to the corresponding values predicted by the single-factor model, we will be able to find estimates of $\lambda_1, \lambda_2, \lambda_3, \psi_1, \psi_2$, and ψ_3 such that the model fits exactly. The six equations derived from the matrix equality implied by the factor analysis model,

$$\mathbf{R} = \begin{pmatrix} \lambda_1 \\ \lambda_2 \\ \lambda_3 \end{pmatrix} \begin{pmatrix} \lambda_1 & \lambda_2 \lambda_3 \end{pmatrix} + \begin{pmatrix} \psi_1 & 0 & 0 \\ 0 & \psi_2 & 0 \\ 0 & 0 & \psi_3 \end{pmatrix},$$

are

$$\hat{\lambda}_1 \lambda_2 = 0.83,$$
$$\hat{\lambda}_1 \lambda_3 = 0.78,$$
$$\hat{\lambda}_1 \lambda_4 = 0.67,$$
$$\psi_1 = 1.0 - \hat{\lambda}_1^2,$$
$$\psi_2 = 1.0 - \hat{\lambda}_2^2,$$
$$\psi_3 = 1.0 - \hat{\lambda}_3^2.$$

The solutions of these equations are

$$\hat{\lambda}_1 = 0.99, \quad \hat{\lambda}_2 = 0.84, \quad \hat{\lambda}_3 = 0.79,$$
$$\hat{\psi}_1 = 0.02, \quad \hat{\psi}_2 = 0.30, \quad \hat{\psi}_3 = 0.38.$$

Suppose now that the observed correlations had been

$$\mathbf{R} = \begin{matrix} \text{Classics} \\ \text{French} \\ \text{English} \end{matrix} \begin{pmatrix} 1.00 & & \\ 0.84 & 1.00 & \\ 0.60 & 0.35 & 1.00 \end{pmatrix}.$$

In this case, the solution for the parameters of a single-factor model is

$$\hat{\lambda}_1 = 1.2, \quad \hat{\lambda}_2 = 0.7, \quad \hat{\lambda}_3 = 0.5,$$
$$\hat{\psi}_1 = -0.44, \quad \hat{\psi}_2 = 0.51, \quad \hat{\psi}_3 = 0.75.$$

Clearly this solution is unacceptable because of the negative estimate for the first specific variance.

In the simple example considered above, the factor analysis model does not give a useful description of the data because the number of parameters in the model equals the number of independent elements in the correlation matrix. In practise, where the k-factor model has fewer parameters than there are independent elements of the covariance or correlation matrix (see Section 5.6), the fitted model represents a genuinely parsimonious description of the data and methods of estimation are needed that try to make the covariance matrix predicted by the factor model as close as possible in some sense to the observed covariance matrix of the manifest variables. There are two main methods of estimation leading to what are known as *principal factor analysis* and *maximum likelihood factor analysis*, both of which are now briefly described.

5.5.1 Principal factor analysis

Principal factor analysis is an eigenvalue and eigenvector technique similar in many respects to principal components analysis (see Chapter 3) but operating not directly on $\mathbf{S}$ (or $\mathbf{R}$) but on what is known as the *reduced covariance matrix*, $\mathbf{S}^*$, defined as

$$\mathbf{S}^* = \mathbf{S} - \hat{\boldsymbol{\Psi}},$$

where $\hat{\boldsymbol{\Psi}}$ is a diagonal matrix containing estimates of the ψ_i. The "ones" on the diagonal of $\mathbf{S}$ have in $\mathbf{S}^*$ been replaced by the estimated communalities, $\sum_{j=1}^k \hat{\lambda}_{ij}^2$, the parts of the variance of each observed variable that can be explained by the common factors. Unlike principal components analysis, factor analysis does not try to account for *all* the observed variance, only that shared through the common factors. Of more concern in factor analysis is accounting for the covariances or correlations between the manifest variables.

To calculate $\mathbf{S}^*$ (or with $\mathbf{R}$ replacing $\mathbf{S}$, $\mathbf{R}^*$) we need values for the communalities. Clearly we cannot calculate them on the basis of factor loadings because these loadings still have to be estimated. To get around this seemingly "chicken and egg" situation, we need to find a sensible way of finding initial values for the communalities that does not depend on knowing the factor loadings. When the factor analysis is based on the correlation matrix of the manifest variables, two frequently used methods are:

- Take the communality of a variable x_i as the square of the multiple correlation coefficient of x_i with the other observed variables.
- Take the communality of x_i as the largest of the absolute values of the correlation coefficients between x_i and one of the other variables.

Each of these possibilities will lead to higher values for the initial communality when x_i is highly correlated with at least some of the other manifest variables, which is essentially what is required.

Given the initial communality values, a principal components analysis is performed on $\mathbf{S}^*$ and the first k eigenvectors used to provide the estimates of the loadings in the k-factor model. The estimation process can stop here or the loadings obtained at this stage can provide revised communality estimates calculated as $\sum_{j=1}^{k} \hat{\lambda}_{ij}^{2}$, where the $\hat{\lambda}_{ij}^{2}$s are the loadings estimated in the previous step. The procedure is then repeated until some convergence criterion is satisfied. Difficulties can sometimes arise with this iterative approach if at any time a communality estimate exceeds the variance of the corresponding manifest variable, resulting in a negative estimate of the variable's specific variance. Such a result is known as a *Heywood case* (see Heywood 1931) and is clearly unacceptable since we cannot have a negative specific variance.

5.5.2 Maximum likelihood factor analysis

Maximum likelihood is regarded, by statisticians at least, as perhaps the most respectable method of estimating the parameters in the factor analysis. The essence of this approach is to assume that the data being analysed have a multivariate normal distribution (see Chapter 1). Under this assumption and assuming the factor analysis model holds, the likelihood function L can be shown to be $-\frac{1}{2}nF$ plus a function of the observations where F is given by

$$F = \ln|\boldsymbol{\Lambda}\boldsymbol{\Lambda}^{\top} + \boldsymbol{\Psi}| + \operatorname{trace}(\mathbf{S}|\boldsymbol{\Lambda}\boldsymbol{\Lambda}^{\top} + \boldsymbol{\Psi}|^{-1}) - \ln|\mathbf{S}| - q.$$

The function F takes the value zero if $\boldsymbol{\Lambda}\boldsymbol{\Lambda}^{\top} + \boldsymbol{\Psi}$ is equal to $\mathbf{S}$ and values greater than zero otherwise. Estimates of the loadings and the specific variances are found by minimising F with respect to these parameters. A number of iterative numerical algorithms have been suggested; for details see Lawley and Maxwell (1963), Mardia et al. (1979), Everitt (1984, 1987), and Rubin and Thayer (1982).

Initial values of the factor loadings and specific variances can be found in a number of ways, including that described above in Section 5.5.1. As with iterated principal factor analysis, the maximum likelihood approach can also experience difficulties with Heywood cases.

5.6 Estimating the number of factors

The decision over how many factors, k, are needed to give an adequate representation of the observed covariances or correlations is generally critical when fitting an exploratory factor analysis model. Solutions with $k = m$ and $k = m + 1$ will often produce quite different factor loadings for *all* factors, unlike a principal components analysis, in which the first m components will be identical in each solution. And, as pointed out by Jolliffe (2002), with too few factors there will be too many high loadings, and with too many factors, factors may be fragmented and difficult to interpret convincingly.

Choosing k might be done by examining solutions corresponding to different values of k and deciding subjectively which can be given the most convincing interpretation. Another possibility is to use the scree diagram approach described in Chapter 3, although the usefulness of this method is not so clear in factor analysis since the eigenvalues represent variances of principal components, not factors.

An advantage of the maximum likelihood approach is that it has an associated formal hypothesis testing procedure that provides a test of the hypothesis H_k that k common factors are sufficient to describe the data against the alternative that the population covariance matrix of the data has no constraints. The test statistic is

$$U = N \min(F),$$

where $N = n + 1 - \frac{1}{6}(2q + 5) - \frac{2}{3}k$. If k common factors are adequate to account for the observed covariances or correlations of the manifest variables (i.e., H_k is true), then U has, asymptotically, a chi-squared distribution with ν degrees of freedom, where

$$\nu = \frac{1}{2}(q - k)^2 - \frac{1}{2}(q + k).$$

In most exploratory studies, k cannot be specified in advance and so a sequential procedure is used. Starting with some small value for k (usually $k = 1$), the parameters in the corresponding factor analysis model are estimated using maximum likelihood. If U is not significant, the current value of k is accepted; otherwise k is increased by one and the process is repeated. If at any stage the degrees of freedom of the test become zero, then either no non-trivial solution is appropriate or alternatively the factor model itself, with its assumption of linearity between observed and latent variables, is questionable. (This procedure is open to criticism because the critical values of the test criterion have not been adjusted to allow for the fact that a set of hypotheses are being tested in sequence.)

5.7 Factor rotation

Up until now, we have conveniently ignored one problematic feature of the factor analysis model, namely that, as formulated in Section 5.3, there is no *unique* solution for the factor loading matrix. We can see that this is so by introducing an orthogonal matrix $\mathbf{M}$ of order $k \times k$ and rewriting the basic regression equation linking the observed and latent variables as

$$\mathbf{x} = (\boldsymbol{\Lambda}\mathbf{M})(\mathbf{M}^\top \mathbf{f}) + \mathbf{u}.$$

This "new" model satisfies all the requirements of a k-factor model as previously outlined with new factors $\mathbf{f}^* = \mathbf{M}\mathbf{f}$ and the new factor loadings $\boldsymbol{\Lambda}\mathbf{M}$. This model implies that the covariance matrix of the observed variables is

$$\boldsymbol{\Sigma} = (\boldsymbol{\Lambda}\mathbf{M})(\boldsymbol{\Lambda}\mathbf{M})^\top + \boldsymbol{\Psi},$$

which, since $\mathbf{MM}^\top = \mathbf{I}$, reduces to $\boldsymbol{\Sigma} = \boldsymbol{\Lambda\Lambda}^\top + \boldsymbol{\Psi}$ as before. Consequently, factors $\mathbf{f}$ with loadings $\boldsymbol{\Lambda}$ and factors $\mathbf{f}^*$ with loadings $\boldsymbol{\Lambda}\mathbf{M}$ are, for any orthogonal matrix $\mathbf{M}$, equivalent for explaining the covariance matrix of the observed variables. Essentially then there are an infinite number of solutions to the factor analysis model as previously formulated.

The problem is generally solved by introducing some constraints in the original model. One possibility is to require the matrix $\mathbf{G}$ given by

$$\mathbf{G} = \boldsymbol{\Lambda\Psi}^{-1}\boldsymbol{\Lambda}$$

to be diagonal, with its elements arranged in descending order of magnitude. Such a requirement sets the first factor to have maximal contribution to the common variance of the observed variables, and the second has maximal contribution to this variance subject to being uncorrelated with the first and so on (cf. principal components analysis in Chapter 3). The constraint above ensures that $\boldsymbol{\Lambda}$ is uniquely determined, except for a possible change of sign of the columns. (When $k = 1$, the constraint is irrelevant.)

The constraints on the factor loadings imposed by a condition such as that given above need to be introduced to make the parameter estimates in the factor analysis model unique, and they lead to orthogonal factors that are arranged in descending order of importance. These properties are not, however, inherent in the factor model, and merely considering such a solution may lead to difficulties of interpretation. For example, two consequences of a factor solution found when applying the constraint above are:

- The factorial complexity of variables is likely to be greater than one regardless of the underlying true model; consequently variables may have substantial loadings on more than one factor.
- Except for the first factor, the remaining factors are often *bipolar*; i.e., they have a mixture of positive and negative loadings.

It may be that a more interpretable orthogonal solution can be achieved using the equivalent model with loadings $\boldsymbol{\Lambda}^* = \boldsymbol{\Lambda}\mathbf{M}$ for some particular orthogonal matrix, $\mathbf{M}$. Such a process is generally known as *factor rotation*, but before we consider how to choose $\mathbf{M}$ (i.e., how to "rotate" the factors), we need to address the question "is factor rotation an acceptable process?"

Certainly factor analysis has in the past been the subject of severe criticism because of the possibility of rotating factors. Critics have suggested that this apparently allows investigators to impose on the data whatever type of solution they are looking for; some have even gone so far as to suggest that factor analysis has become popular in some areas precisely because it *does* enable users to impose their preconceived ideas of the structure behind the observed correlations (Blackith and Reyment 1971). But, on the whole, such suspicions are not justified and factor rotation can be a useful procedure for simplifying an exploratory factor analysis. Factor rotation merely allows the fitted factor analysis model to be described as simply as possible; rotation does not alter the overall structure of a solution but only how the solution is described. Rotation is a process by which a solution is made more interpretable without changing its underlying mathematical properties. Initial factor solutions with variables loading on several factors and with bipolar factors can be difficult to interpret. Interpretation is more straightforward if each variable is highly loaded on at most one factor and if all factor loadings are either large and positive or near zero, with few intermediate values. The variables are thus split into disjoint sets, each of which is associated with a single factor. This aim is essentially what Thurstone (1931) referred to as *simple structure*. In more detail, such structure has the following properties:

- Each row or the factor loading matrix should contain at least one zero.
- Each column of the loading matrix should contain at least k zeros.
- Every pair of columns of the loading matrix should contain several variables whose loadings vanish in one column but not in the other.
- If the number of factors is four or more, every pair of columns should contain a large number of variables with zero loadings in both columns.
- Conversely, for every pair of columns of the loading matrix only a small number of variables should have non-zero loadings in both columns.

When simple structure is achieved, the observed variables will fall into mutually exclusive groups whose loadings are high on single factors, perhaps moderate to low on a few factors, and of negligible size on the remaining factors. Medium-sized, equivocal loadings are to be avoided.

The search for simple structure or something close to it begins after an initial factoring has determined the number of common factors necessary and the communalities of each observed variable. The factor loadings are then transformed by post-multiplication by a suitably chosen orthogonal matrix. Such a transformation is equivalent to a rigid rotation of the axes of the originally identified factor space. And during the rotation phase of the analysis, we might choose to abandon one of the assumptions made previously, namely that factors are orthogonal, i.e., independent (the condition was assumed initially simply for convenience in describing the factor analysis model). Consequently, two types of rotation are possible:

- *orthogonal rotation*, in which methods restrict the rotated factors to being uncorrelated, or
- *oblique rotation*, where methods allow correlated factors.

As we have seen above, orthogonal rotation is achieved by post-multiplying the original matrix of loadings by an orthogonal matrix. For oblique rotation, the original loadings matrix is post-multiplied by a matrix that is no longer constrained to be orthogonal. With an orthogonal rotation, the matrix of correlations between factors after rotation is the identity matrix. With an oblique rotation, the corresponding matrix of correlations is restricted to have unit elements on its diagonal, but there are no restrictions on the off-diagonal elements.

So the first question that needs to be considered when rotating factors is whether we should use an orthogonal or an oblique rotation. As for many questions posed in data analysis, there is no universal answer to this question. There are advantages and disadvantages to using either type of rotation procedure. As a general rule, if a researcher is primarily concerned with getting results that "best fit" his or her data, then the factors should be rotated obliquely. If, on the other hand, the researcher is more interested in the generalisability of his or her results, then orthogonal rotation is probably to be preferred.

One major advantage of an orthogonal rotation is simplicity since the loadings represent correlations between factors and manifest variables. This is *not* the case with an oblique rotation because of the correlations between the factors. Here there are two parts of the solution to consider;

- *factor pattern coefficients*, which are regression coefficients that multiply with factors to produce measured variables according to the common factor model, and
- *factor structure coefficients*, correlation coefficients between manifest variables and the factors.

Additionally there is a matrix of factor correlations to consider. In many cases where these correlations are relatively small, researchers may prefer to return to an orthogonal solution.

There are a variety of rotation techniques, although only relatively few are in general use. For orthogonal rotation, the two most commonly used techniques are known as *varimax* and *quartimax*.

- *Varimax rotation*, originally proposed by Kaiser (1958), has as its rationale the aim of factors with a few large loadings and as many near-zero loadings as possible. This is achieved by iterative maximisation of a quadratic function of the loadings–details are given in Mardia et al. (1979). It produces factors that have high correlations with one small set of variables and little or no correlation with other sets. There is a tendency for any general factor to disappear because the factor variance is redistributed.
- *Quartimax rotation*, originally suggested by Carroll (1953), forces a given variable to correlate highly on one factor and either not at all or very low on other factors. It is far less popular than varimax.

For oblique rotation, the two methods most often used are *oblimin* and *promax*.

- *Oblimin rotation*, invented by Jennrich and Sampson (1966), attempts to find simple structure with regard to the factor pattern matrix through a parameter that is used to control the degree of correlation between the factors. Fixing a value for this parameter is not straightforward, but Pett, Lackey, and Sullivan (2003) suggest that values between about −0.5 and 0.5 are sensible for many applications.

- *Promax rotation*, a method due to Hendrickson and White (1964), operates by raising the loadings in an orthogonal solution (generally a varimax rotation) to some power. The goal is to obtain a solution that provides the best structure using the lowest possible power loadings and the lowest correlation between the factors.

Factor rotation is often regarded as controversial since it apparently allows the investigator to impose on the data whatever type of solution is required. But this is clearly *not* the case since although the axes may be rotated about their origin or may be allowed to become oblique, *the distribution of the points will remain invariant*. Rotation is simply a procedure that allows new axes to be chosen so that the positions of the points can be described as simply as possible.

(It should be noted that rotation techniques are also often applied to the results from a principal components analysis in the hope that they will aid in their interpretability. Although in some cases this may be acceptable, it does have several disadvantages, which are listed by Jolliffe (1989). The main problem is that the defining property of principal components, namely that of accounting for maximal proportions of the total variation in the observed variables, is lost after rotation.

5.8 Estimating factor scores

The first stage of an exploratory factor analysis consists of the estimation of the parameters in the model and the rotation of the factors, followed by an (often heroic) attempt to interpret the fitted model. The second stage is concerned with estimating latent variable scores for each individual in the data set; such factor scores are often useful for a number of reasons:

1. They represent a parsimonious summary of the original data possibly useful in subsequent analyses (cf. principal component scores in Chapter 3).
2. They are likely to be more reliable than the observed variable values.
3. The factor score is a "pure" measure of a latent variable, while an observed value may be ambiguous because we do not know what combination of latent variables may be represented by that observed value.

But the calculation of factor scores is not as straightforward as the calculation of principal component scores. In the original equation defining the factor analysis model, the variables are expressed in terms of the factors, whereas to calculate scores we require the relationship to be in the opposite direction. Bartholomew and Knott (1987) make the point that to talk about "estimating" factor scores is essentially misleading since they are random variables and the issue is really one of prediction. But if we make the assumption of normality, the conditional distribution of $\mathbf{f}$ given $\mathbf{x}$ can be found. It is

$$N(\boldsymbol{\Lambda}^\top \boldsymbol{\Sigma}^{-1} \mathbf{x}, (\boldsymbol{\Lambda}^\top \boldsymbol{\Psi}^{-1} \boldsymbol{\Lambda} + \mathbf{I})^{-1}).$$

Consequently, one plausible way of calculating factor scores would be to use the sample version of the mean of this distribution, namely

$$\hat{\mathbf{f}} = \hat{\boldsymbol{\Lambda}}^{\top}\mathbf{S}^{-1}\mathbf{x},$$

where the vector of scores for an individual, $\mathbf{x}$, is assumed to have mean zero; i.e., sample means for each variable have already been subtracted. Other possible methods for deriving factor scores are described in Rencher (1995), and helpful detailed calculations of several types of factor scores are given in Hershberger (2005). In many respects, the most damaging problem with factor analysis is not the rotational indeterminacy of the loadings but the indeterminacy of the factor scores.

5.9 Two examples of exploratory factor analysis

5.9.1 Expectations of life

The data in Table 5.1 show life expectancy in years by country, age, and sex. The data come from Keyfitz and Flieger (1971) and relate to life expectancies in the 1960s.

Table 5.1: life data. Life expectancies for different countries by age and gender.

	m0	m25	m50	m75	w0	w25	w50	w75
Algeria	63	51	30	13	67	54	34	15
Cameroon	34	29	13	5	38	32	17	6
Madagascar	38	30	17	7	38	34	20	7
Mauritius	59	42	20	6	64	46	25	8
Reunion	56	38	18	7	62	46	25	10
Seychelles	62	44	24	7	69	50	28	14
South Africa (C)	50	39	20	7	55	43	23	8
South Africa (W)	65	44	22	7	72	50	27	9
Tunisia	56	46	24	11	63	54	33	19
Canada	69	47	24	8	75	53	29	10
Costa Rica	65	48	26	9	68	50	27	10
Dominican Rep.	64	50	28	11	66	51	29	11
El Salvador	56	44	25	10	61	48	27	12
Greenland	60	44	22	6	65	45	25	9
Grenada	61	45	22	8	65	49	27	10
Guatemala	49	40	22	9	51	41	23	8
Honduras	59	42	22	6	61	43	22	7
Jamaica	63	44	23	8	67	48	26	9
Mexico	59	44	24	8	63	46	25	8

Table 5.1: `life` data (continued).

	m0	m25	m50	m75	w0	w25	w50	w75
Nicaragua	65	48	28	14	68	51	29	13
Panama	65	48	26	9	67	49	27	10
Trinidad (62)	64	63	21	7	68	47	25	9
Trinidad (67)	64	43	21	6	68	47	24	8
United States (66)	67	45	23	8	74	51	28	10
United States (NW66)	61	40	21	10	67	46	25	11
United States (W66)	68	46	23	8	75	52	29	10
United States (67)	67	45	23	8	74	51	28	10
Argentina	65	46	24	9	71	51	28	10
Chile	59	43	23	10	66	49	27	12
Colombia	58	44	24	9	62	47	25	10
Ecuador	57	46	28	9	60	49	28	11

To begin, we will use the formal test for the number of factors incorporated into the maximum likelihood approach. We can apply this test to the data, assumed to be contained in the data frame `life` with the country names labelling the rows and variable names as given in Table 5.1, using the following R code:

```
R> sapply(1:3, function(f)
+       factanal(life, factors = f, method ="mle")$PVAL)

objective objective objective
1.880e-24 1.912e-05 4.578e-01
```

These results suggest that a three-factor solution might be adequate to account for the observed covariances in the data, although it has to be remembered that, with only 31 countries, use of an asymptotic test result may be rather suspect. The three-factor solution is as follows (note that the solution is that resulting from a varimax solution. the default for the `factanal()` function):

```
R> factanal(life, factors = 3, method ="mle")

Call:
factanal(x = life, factors = 3, method = "mle")

Uniquenesses:
   m0    m25    m50    m75     w0    w25    w50    w75
0.005  0.362  0.066  0.288  0.005  0.011  0.020  0.146

Loadings:
     Factor1 Factor2 Factor3
```

```
m0   0.964   0.122   0.226
m25  0.646   0.169   0.438
m50  0.430   0.354   0.790
m75          0.525   0.656
w0   0.970   0.217
w25  0.764   0.556   0.310
w50  0.536   0.729   0.401
w75  0.156   0.867   0.280
```

	Factor1	Factor2	Factor3
SS loadings	3.375	2.082	1.640
Proportion Var	0.422	0.260	0.205
Cumulative Var	0.422	0.682	0.887

```
Test of the hypothesis that 3 factors are sufficient.
The chi square statistic is 6.73 on 7 degrees of freedom.
The p-value is 0.458
```

("Blanks" replace negligible loadings.) Examining the estimated factor loadings, we see that the first factor is dominated by life expectancy at birth for both males and females; perhaps this factor could be labelled "life force at birth". The second reflects life expectancies at older ages, and we might label it "life force amongst the elderly". The third factor from the varimax rotation has its highest loadings for the life expectancies of men aged 50 and 75 and in the same vein might be labelled "life force for elderly men". (When labelling factors in this way, factor analysts can often be extremely creative!)

The estimated factor scores are found as follows;

```
R> (scores <- factanal(life, factors = 3, method = "mle",
+                     scores = "regression")$scores)
```

	Factor1	Factor2	Factor3
Algeria	-0.258063	1.90096	1.91582
Cameroon	-2.782496	-0.72340	-1.84772
Madagascar	-2.806428	-0.81159	-0.01210
Mauritius	0.141005	-0.29028	-0.85862
Reunion	-0.196352	0.47430	-1.55046
Seychelles	0.367371	0.82902	-0.55214
South Africa (C)	-1.028568	-0.08066	-0.65422
South Africa (W)	0.946194	0.06400	-0.91995
Tunisia	-0.862494	3.59177	-0.36442
Canada	1.245304	0.29564	-0.27343
Costa Rica	0.508736	-0.50500	1.01329
Dominican Rep.	0.106044	0.01111	1.83872
El Salvador	-0.608156	0.65101	0.48836
Greenland	0.235114	-0.69124	-0.38559

Grenada	0.132008	0.25241	-0.15221
Guatemala	-1.450336	-0.67766	0.65912
Honduras	0.043253	-1.85176	0.30633
Jamaica	0.462125	-0.51918	0.08033
Mexico	-0.052333	-0.72020	0.44418
Nicaragua	0.268974	0.08407	1.70568
Panama	0.442333	-0.73778	1.25219
Trinidad (62)	0.711367	-0.95989	-0.21545
Trinidad (67)	0.787286	-1.10729	-0.51958
United States (66)	1.128331	0.16390	-0.68177
United States (NW66)	0.400059	-0.36230	-0.74299
United States (W66)	1.214345	0.40877	-0.69225
United States (67)	1.128331	0.16390	-0.68177
Argentina	0.731345	0.24812	-0.12818
Chile	0.009752	0.75223	-0.49199
Colombia	-0.240603	-0.29544	0.42920
Ecuador	-0.723452	0.44246	1.59165

We can use the scores to provide the plot of the data shown in Figure 5.1.

Ordering along the first axis reflects life force at birth ranging from Cameroon and Madagascar to countries such as the USA. And on the third axis Algeria is prominent because it has high life expectancy amongst men at higher ages, with Cameroon at the lower end of the scale with a low life expectancy for men over 50.

5.9.2 Drug use by American college students

The majority of adult and adolescent Americans regularly use psychoactive substances during an increasing proportion of their lifetimes. Various forms of licit and illicit psychoactive substance use are prevalent, suggesting that patterns of psychoactive substance taking are a major part of the individual's behavioural repertoire and have pervasive implications for the performance of other behaviours. In an investigation of these phenomena, Huba, Wingard, and Bentler (1981) collected data on drug usage rates for 1634 students in the seventh to ninth grades in 11 schools in the greater metropolitan area of Los Angeles. Each participant completed a questionnaire about the number of times a particular substance had ever been used. The substances asked about were as follows:

- cigarettes;
- beer;
- wine;
- liquor;
- cocaine;
- tranquillizers;
- drug store medications used to get high;

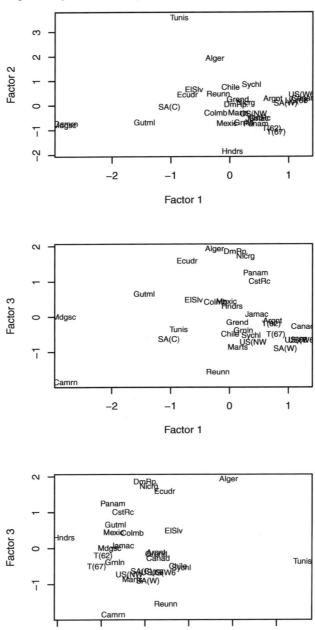

Fig. 5.1. Individual scatterplots of three factor scores for life expectancy data, with points labelled by abbreviated country names.

- heroin and other opiates;
- marijuana;
- hashish;
- inhalants (glue, gasoline, etc.);
- hallucinogenics (LSD, mescaline, etc.);
- amphetamine stimulants.

Responses were recorded on a five-point scale: never tried, only once, a few times, many times, and regularly. The correlations between the usage rates of the 13 substances are shown in Figure 5.2. The plot was produced using the `levelplot()` function from the package **lattice** with a somewhat lengthy panel function, so we refer the interested reader to the R code contained in the demo for this chapter (see the Preface for how to access this document). The figure depicts each correlation by an ellipse whose shape tends towards a line with slope 1 for correlations near 1, to a circle for correlations near zero, and to a line with negative slope -1 for negative correlations near -1. In addition, 100 times the correlation coefficient is printed inside the ellipse, and a colourcoding indicates strong negative (dark) to strong positive (light) correlations.

We first try to determine the number of factors using the maximum likelihood test. The R code for finding the results of the test for number of factors here is:

```
R> sapply(1:6, function(nf)
+       factanal(covmat = druguse, factors = nf,
+               method = "mle", n.obs = 1634)$PVAL)
```

```
objective objective objective objective objective objective
0.000e+00 9.786e-70 7.364e-28 1.795e-11 3.892e-06 9.753e-02
```

These values suggest that only the six-factor solution provides an adequate fit. The results from the six-factor varimax solution are obtained from

```
R> (factanal(covmat = druguse, factors = 6,
+           method = "mle", n.obs = 1634))
```

```
Call:
factanal(factors = 6, covmat = druguse, n.obs = 1634)
```

```
Uniquenesses:
            cigarettes                    beer
                 0.563                   0.368
                  wine                  liquor
                 0.374                   0.412
               cocaine          tranquillizers
                 0.681                   0.522
  drug store medication                  heroin
                 0.785                   0.669
```

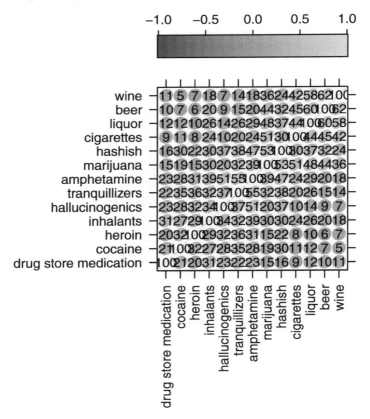

Fig. 5.2. Visualisation of the correlation matrix of drug use. The numbers in the cells correspond to 100 times the correlation coefficient. The color and the shape of the plotting symbols also correspond to the correlation in this cell.

```
        marijuana                    hashish
           0.318                      0.005
        inhalants              hallucinogenics
           0.541                      0.620
      amphetamine
           0.005
```

```
Loadings:
             Factor1 Factor2 Factor3 Factor4 Factor5
cigarettes    0.494                           0.407
beer          0.776                           0.112
wine          0.786
liquor        0.720   0.121   0.103   0.115   0.160
```

	Factor1	Factor2	Factor3	Factor4	Factor5
cocaine		0.519		0.132	
tranquillizers	0.130	0.564	0.321	0.105	0.143
drug store medication		0.255			
heroin		0.532	0.101		
marijuana	0.429	0.158	0.152	0.259	0.609
hashish	0.244	0.276	0.186	0.881	0.194
inhalants	0.166	0.308	0.150		0.140
hallucinogenics		0.387	0.335	0.186	
amphetamine	0.151	0.336	0.886	0.145	0.137

	Factor6
cigarettes	0.110
beer	
wine	
liquor	
cocaine	0.158
tranquillizers	
drug store medication	0.372
heroin	0.190
marijuana	0.110
hashish	0.100
inhalants	0.537
hallucinogenics	0.288
amphetamine	0.187

	Factor1	Factor2	Factor3	Factor4	Factor5	Factor6
SS loadings	2.301	1.415	1.116	0.964	0.676	0.666
Proportion Var	0.177	0.109	0.086	0.074	0.052	0.051
Cumulative Var	0.177	0.286	0.372	0.446	0.498	0.549

Test of the hypothesis that 6 factors are sufficient.
The chi square statistic is 22.41 on 15 degrees of freedom.
The p-value is 0.0975

Substances that load highly on the first factor are cigarettes, beer, wine, liquor, and marijuana and we might label it "social/soft drug use". Cocaine, tranquillizers, and heroin load highly on the second factor–the obvious label for the factor is "hard drug use". Factor three is essentially simply amphetamine use, and factor four hashish use. We will not try to interpret the last two factors, even though the formal test for number of factors indicated that a six-factor solution was necessary. It may be that we should not take the results of the formal test too literally; rather, it may be a better strategy to consider the value of k indicated by the test to be an upper bound on the number of factors with practical importance. Certainly a six-factor solution for a data set with only 13 manifest variables might be regarded as not entirely satisfactory, and clearly we would have some difficulties interpreting all the factors.

One of the problems is that with the large sample size in this example, even small discrepancies between the correlation matrix predicted by a proposed model and the observed correlation matrix may lead to rejection of the model. One way to investigate this possibility is simply to look at the differences between the observed and predicted correlations. We shall do this first for the six-factor model using the following R code:

```
R> pfun <- function(nf) {
+        fa <- factanal(covmat = druguse, factors = nf,
+                       method = "mle", n.obs = 1634)
+        est <- tcrossprod(fa$loadings) + diag(fa$uniquenesses)
+        ret <- round(druguse - est, 3)
+        colnames(ret) <- rownames(ret) <-
+            abbreviate(rownames(ret), 3)
+        ret
+ }
R> pfun(6)
```

	cgr	ber	win	lqr	ccn	trn	dsm	hrn
cgr	0.000	-0.001	0.014	-0.018	0.010	0.001	-0.020	-0.004
ber	-0.001	0.000	-0.002	0.004	0.004	-0.011	-0.001	0.007
win	0.014	-0.002	0.000	-0.001	-0.001	-0.005	0.008	0.008
lqr	-0.018	0.004	-0.001	0.000	-0.008	0.021	-0.006	-0.018
ccn	0.010	0.004	-0.001	-0.008	0.000	0.000	0.008	0.004
trn	0.001	-0.011	-0.005	0.021	0.000	0.000	0.006	-0.004
dsm	-0.020	-0.001	0.008	-0.006	0.008	0.006	0.000	-0.015
hrn	-0.004	0.007	0.008	-0.018	0.004	-0.004	-0.015	0.000
mrj	0.001	0.002	-0.004	0.003	-0.004	-0.004	0.008	0.006
hsh	0.000	0.000	0.000	0.000	0.000	0.000	0.000	0.000
inh	0.010	-0.004	-0.007	0.012	-0.003	0.002	0.004	-0.002
hll	-0.005	0.005	-0.001	-0.005	-0.008	-0.008	-0.002	0.020
amp	0.000	0.000	0.000	0.000	0.000	0.000	0.000	0.000

	mrj	hsh	inh	hll	amp
cgr	0.001	0	0.010	-0.005	0
ber	0.002	0	-0.004	0.005	0
win	-0.004	0	-0.007	-0.001	0
lqr	0.003	0	0.012	-0.005	0
ccn	-0.004	0	-0.003	-0.008	0
trn	-0.004	0	0.002	-0.008	0
dsm	0.008	0	0.004	-0.002	0
hrn	0.006	0	-0.002	0.020	0
mrj	0.000	0	-0.006	0.003	0
hsh	0.000	0	0.000	0.000	0
inh	-0.006	0	0.000	-0.002	0
hll	0.003	0	-0.002	0.000	0
amp	0.000	0	0.000	0.000	0

The differences are all very small, underlining that the six-factor model does describe the data very well. Now let us look at the corresponding matrices for the three- and four-factor solutions found in a similar way in Figure 5.3. Again, in both cases the residuals are all relatively small, suggesting perhaps that use of the formal test for number of factors leads, in this case, to overfitting. The three-factor model appears to provide a perfectly adequate fit for these data.

5.10 Factor analysis and principal components analysis compared

Factor analysis, like principal components analysis, is an attempt to explain a set of multivariate data using a smaller number of dimensions than one begins with, but the procedures used to achieve this goal are essentially quite different in the two approaches. Some differences between the two are as follows:

- Factor analysis tries to explain the covariances or correlations of the observed variables by means of a few common factors. Principal components analysis is primarily concerned with explaining the variance of the observed variables.
- If the number of retained components is increased, say from m to $m+1$, the first m components are unchanged. This is not the case in factor analysis, where there can be substantial changes in *all* factors if the number of factors is changed.
- The calculation of principal component scores is straightforward, but the calculation of factor scores is more complex, and a variety of methods have been suggested.
- There is usually no relationship between the principal components of the sample correlation matrix and the sample covariance matrix. For maximum likelihood factor analysis, however, the results of analysing either matrix are essentially equivalent (which is not true of principal factor analysis).

Despite these differences, the results from both types of analyses are frequently very similar. Certainly, if the specific variances are small, we would expect both forms of analyses to give similar results. However, if the specific variances are large, they will be absorbed into all the principal components, both retained and rejected, whereas factor analysis makes special provision for them.

Lastly, it should be remembered that both principal components analysis and factor analysis are similar in one important respect–they are both pointless if the observed variables are almost uncorrelated. In this case, factor analysis has nothing to explain and principal components analysis will simply lead to components that are similar to the original variables.

```
R> pfun(3)
        cgr    ber    win    lqr    ccn    trn    dsm    hrn    mrj    hsh    inh    hll    amp
cgr   0.000 -0.001  0.009 -0.013  0.011  0.009 -0.011 -0.004  0.003 -0.027  0.039 -0.017  0.002
ber  -0.001  0.000 -0.002  0.002  0.002 -0.014  0.000  0.005 -0.001  0.019 -0.002  0.009 -0.007
win   0.009 -0.002  0.000  0.000 -0.002 -0.004  0.012  0.013  0.001 -0.017 -0.007  0.004  0.002
lqr  -0.013  0.002  0.000  0.000 -0.008  0.024 -0.017 -0.020 -0.001  0.014 -0.002 -0.015  0.006
ccn   0.011  0.002 -0.002 -0.008  0.000  0.031  0.038  0.082 -0.002  0.041  0.023 -0.030 -0.075
trn   0.009 -0.014 -0.004  0.024  0.031  0.000 -0.021  0.026  0.007 -0.016 -0.038 -0.058  0.044
dsm  -0.011  0.000  0.012 -0.017  0.038 -0.021  0.000  0.021  0.006 -0.040  0.113  0.000 -0.038
hrn  -0.004  0.005  0.013 -0.020  0.082  0.026  0.021  0.000  0.006 -0.035  0.031 -0.005 -0.049
mrj   0.003 -0.001  0.001 -0.001 -0.002  0.007  0.006  0.006  0.000  0.001  0.003 -0.002 -0.002
hsh  -0.027  0.019 -0.017  0.014  0.041 -0.016 -0.040 -0.035  0.001  0.000 -0.035  0.034  0.010
inh   0.039 -0.002 -0.007 -0.002  0.023 -0.038  0.113  0.031  0.003 -0.035  0.000  0.007 -0.015
hll  -0.017  0.009  0.004 -0.015 -0.030 -0.058  0.000 -0.005 -0.002  0.034  0.007  0.000  0.041
amp   0.002 -0.007  0.002  0.006 -0.075  0.044 -0.038 -0.049 -0.002  0.010 -0.015  0.041  0.000
```

```
R> pfun(4)
        cgr    ber    win    lqr    ccn    trn    dsm    hrn    mrj    hsh    inh    hll    amp
cgr   0.000 -0.001  0.008 -0.012  0.009  0.008 -0.015 -0.007  0.001 -0.023  0.037 -0.020  0.000
ber  -0.001  0.000 -0.001  0.001  0.000 -0.016 -0.002  0.003 -0.001  0.018 -0.005  0.006  0.000
win   0.008 -0.001  0.000  0.000 -0.001 -0.005  0.012  0.014  0.001 -0.020 -0.008  0.001  0.000
lqr  -0.012  0.001  0.000  0.000 -0.004  0.029 -0.015 -0.015 -0.001  0.018  0.001 -0.010 -0.001
ccn   0.009  0.000 -0.001 -0.004  0.000  0.024 -0.014 -0.020 -0.003  0.035 -0.022 -0.028  0.000
trn   0.008 -0.016 -0.005  0.029  0.024  0.000 -0.020  0.027 -0.001  0.001 -0.032 -0.028  0.001
dsm  -0.015 -0.002  0.012 -0.015 -0.014 -0.020  0.000 -0.018  0.003 -0.042  0.090  0.008  0.000
hrn  -0.007  0.003  0.014 -0.015 -0.020  0.027 -0.018  0.000  0.003 -0.037  0.000  0.005  0.000
mrj   0.001 -0.001  0.001 -0.001 -0.003 -0.001  0.003  0.003  0.000  0.000  0.001 -0.002  0.000
hsh  -0.023  0.018 -0.020  0.018  0.035  0.001 -0.042 -0.037  0.000  0.000 -0.031  0.055 -0.001
inh   0.037 -0.005 -0.008  0.001 -0.022 -0.032  0.090  0.000  0.001 -0.031  0.000  0.021  0.000
hll  -0.020  0.006  0.001 -0.010 -0.028 -0.028  0.008  0.005 -0.002  0.055  0.021  0.000  0.000
amp   0.000  0.000  0.000 -0.001  0.000  0.001  0.000  0.000  0.000 -0.001  0.000  0.000  0.000
```

Fig. 5.3. Differences between three- and four-factor solutions and actual correlation matrix for the drug use data.

5.11 Summary

Factor analysis has probably attracted more critical comments than any other statistical technique. Hills (1977), for example, has gone so far as to suggest that factor analysis is not worth the time necessary to understand it and carry it out. And Chatfield and Collins (1980) recommend that factor analysis should not be used in most practical situations. The reasons for such an openly sceptical view about factor analysis arise first from the central role of latent variables in the factor analysis model and second from the lack of uniqueness of the factor loadings in the model, which gives rise to the possibility of rotating factors. It certainly is the case that, since the common factors cannot be measured or observed, the existence of these hypothetical variables is open to question. A factor is a construct operationally defined by its factor loadings, and overly enthusiastic reification is not recommended. And it is the case that, given one factor loading matrix, there are an infinite number of factor loading matrices that could equally well (or equally badly) account for the variances and covariances of the manifest variables. Rotation methods are designed to find an easily interpretable solution from among this infinitely large set of alternatives by finding a solution that exhibits the best simple structure.

Factor analysis can be a useful tool for investigating particular features of the structure of multivariate data. Of course, like many models used in data analysis, the one used in factor analysis may be only a very idealised approximation to the truth. Such an approximation may, however, prove a valuable starting point for further investigations, particularly for the confirmatory factor analysis models that are the subject of Chapter 7.

For exploratory factor analysis, similar comments apply about the size of n and q needed to get convincing results, such as those given in Chapter 3 for principal components analysis. And the maximum likelihood method for the estimation of factor loading and specific variances used in this chapter is only suitable for data having a multivariate normal distribution (or at least a reasonable approximation to such a distribution). Consequently, for the factor analysis of, in particular, binary variables, special methods are needed; see, for example, Muthen (1978).

5.12 Exercises

Ex. 5.1 Show how the result $\mathbf{\Sigma} = \mathbf{\Lambda}\mathbf{\Lambda}^\top + \mathbf{\Psi}$ arises from the assumptions of uncorrelated factors, independence of the specific variates, and independence of common factors and specific variances. What form does $\mathbf{\Sigma}$ take if the factors are allowed to be correlated?

Ex. 5.2 Show that the communalities in a factor analysis model are unaffected by the transformation $\mathbf{\Lambda}^* = \mathbf{\Lambda}\mathbf{M}$.

Ex. 5.3 Give a formula for the proportion of variance explained by the jth factor estimated by the principal factor approach.

Ex. 5.4 Apply the factor analysis model separately to the life expectancies of men and women and compare the results.

Ex. 5.5 The correlation matrix given below arises from the scores of 220 boys in six school subjects: (1) French, (2) English, (3) History, (4) Arithmetic, (5) Algebra, and (6) Geometry. Find the two-factor solution from a maximum likelihood factor analysis. By plotting the derived loadings, find an orthogonal rotation that allows easier interpretation of the results.

$$\mathbf{R} = \begin{array}{l} \text{French} \\ \text{English} \\ \text{History} \\ \text{Arithmetic} \\ \text{Algebra} \\ \text{Geometry} \end{array} \left(\begin{array}{llllll} 1.00 & & & & & \\ 0.44 & 1.00 & & & & \\ 0.41 & 0.35 & 1.00 & & & \\ 0.29 & 0.35 & 0.16 & 1.00 & & \\ 0.33 & 0.32 & 0.19 & 0.59 & 1.00 & \\ 0.25 & 0.33 & 0.18 & 0.47 & 0.46 & 1.00 \end{array} \right) .$$

Ex. 5.6 The matrix below shows the correlations between ratings on nine statements about pain made by 123 people suffering from extreme pain. Each statement was scored on a scale from 1 to 6, ranging from agreement to disagreement. The nine pain statements were as follows:

1. Whether or not I am in pain in the future depends on the skills of the doctors.
2. Whenever I am in pain, it is usually because of something I have done or not done,
3. Whether or not I am in pain depends on what the doctors do for me.
4. I cannot get any help for my pain unless I go to seek medical advice.
5. When I am in pain I know that it is because I have not been taking proper exercise or eating the right food.
6. People's pain results from their own carelessness.
7. I am directly responsible for my pain,
8. relief from pain is chiefly controlled by the doctors.
9. People who are never in pain are just plain lucky.

$$\left(\begin{array}{rrrrrrrrr} 1.00 & & & & & & & & \\ -0.04 & 1.00 & & & & & & & \\ 0.61 & -0.07 & 1.00 & & & & & & \\ 0.45 & -0.12 & 0.59 & 1.00 & & & & & \\ 0.03 & 0.49 & 0.03 & -0.08 & 1.00 & & & & \\ -0.29 & 0.43 & -0.13 & -0.21 & 0.47 & 1.00 & & & \\ -0.30 & 0.30 & -0.24 & -0.19 & 0.41 & 0.63 & 1.00 & & \\ 0.45 & -0.31 & 0.59 & 0.63 & -0.14 & -0.13 & -0.26 & 1.00 & \\ 0.30 & -0.17 & 0.32 & 0.37 & -0.24 & -0.15 & -0.29 & 0.40 & 1.00 \end{array} \right) .$$

(a) Perform a principal components analysis on these data, and examine the associated scree plot to decide on the appropriate number of components.

(b) Apply maximum likelihood factor analysis, and use the test described in the chapter to select the necessary number of common factors.

(c) Rotate the factor solution selected using both an orthogonal and an oblique procedure, and interpret the results.

6

Cluster Analysis

6.1 Introduction

> An intelligent being cannot treat every object it sees as a unique entity unlike anything else in the universe. It has to put objects in categories so that it may apply its hard-won knowledge about similar objects encountered in the past to the object at hand (Pinker 1997).

One of the most basic abilities of living creatures involves the grouping of similar objects to produce a classification. The idea of sorting similar things into categories is clearly a primitive one because early humans, for example, must have been able to realise that many individual objects shared certain properties such as being edible, or poisonous, or ferocious, and so on. And classification in its widest sense is needed for the development of language, which consists of words that help us to recognise and discuss the different types of events, objects, and people we encounter. Each noun in a language, for example, is essentially a label used to describe a class of things that have striking features in common; thus animals are called cats, dogs, horses, etc., and each name collects individuals into groups. Naming and classifying are essentially synonymous.

As well as being a basic human conceptual activity, classification of the phenomena being studied is an important component of virtually all scientific research. In the behavioural sciences, for example, these "phenomena" may be individuals or societies, or even patterns of behaviour or perception. The investigator is usually interested in finding a classification in which the items of interest are sorted into a small number of *homogeneous groups* or *clusters*, the terms being synonymous. Most commonly the required classification is one in which the groups are *mutually exclusive* (an item belongs to a single group) rather than *overlapping* (items can be members of more than one group). At the very least, any derived classification scheme should provide a convenient method of organizing a large, complex set of multivariate data, with the class labels providing a parsimonious way of describing the patterns of similarities and differences in the data. In market research, for example, it might be useful to group a large number of potential customers according to their needs in a particular product area. Advertising campaigns might then be tailored for the different types of consumers as represented by the different groups.

But often a classification may seek to serve a more fundamental purpose. In psychiatry, for example, the classification of psychiatric patients with different symptom profiles into clusters might help in the search for the causes of mental illnesses and perhaps even lead to improved therapeutic methods. And these twin aims of *prediction* (separating diseases that require different treatments) and *aetiology* (searching for the causes of disease) for classifications will be the same in other branches of medicine.

Clearly, a variety of classifications will always be possible for whatever is being classified. Human beings could, for example, be classified with respect to economic status into groups labelled *lower class*, *middle class*, and *upper class* or they might be classified by annual consumption of alcohol into *low*, *medium*, and *high*. Clearly, different classifications may not collect the same set of individuals into groups, but some classifications will be more useful than others, a point made clearly by the following extract from Needham (1965) in which he considers the classification of human beings into men and women:

> The usefulness of this classification does not begin and end with all that can, in one sense, be strictly inferred from it–namely a statement about sexual organs. It is a very useful classification because classifying a person as man or woman conveys a great deal more information, about probable relative size, strength, certain types of dexterity and so on. When we say that persons in class *man* are more suitable than persons in class *woman* for certain tasks and conversely, we are only incidentally making a remark about sex, our primary concern being with strength, endurance, etc. The point is that we have been able to use a classification of persons which conveys information on many properties. On the contrary a classification of persons into those with hairs on their forearms between 3/16 and $\frac{1}{4}$ inch long and those without, though it may serve some particular use, is certainly of no general use, for imputing membership in the former class to a person conveys information on this property alone. Put another way, there are no known properties which divide up a set of people in a similar way.

In a similar vein, a classification of books based on subject matter into classes such as dictionaries, novels, biographies, and so on is likely to be far more useful than one based on, say, the colour of the book's binding. Such examples illustrate that any classification of a set of multivariate data is likely to be judged on its usefulness.

6.2 Cluster analysis

Cluster analysis is a generic term for a wide range of numerical methods with the common goal of uncovering or discovering groups or clusters of observations that are homogeneous and separated from other groups. Clustering techniques essentially try to formalise what human observers do so well in two or three dimensions. Consider, for example, the scatterplot shown in Figure 6.1. The conclusion that there are three natural groups or clusters of dots is reached with no conscious effort or thought. Clusters are identified by the assessment of the relative distances between points, and in this example the relative homogeneity of each cluster and the degree of their separation makes the task very simple. The examination of scatterplots based either on the original data or perhaps on the first few principal component scores of the data is often a very helpful initial phase when intending to apply some form of cluster analysis to a set of multivariate data.

Cluster analysis techniques are described in detail in Gordon (1987, 1999) in and Everitt, Landau, Leese, and Stahl (2011). In this chapter, we give a relatively brief account of three types of clustering methods: *agglomerative hierarchical techniques*, *k-means clustering*, and *model-based clustering*.

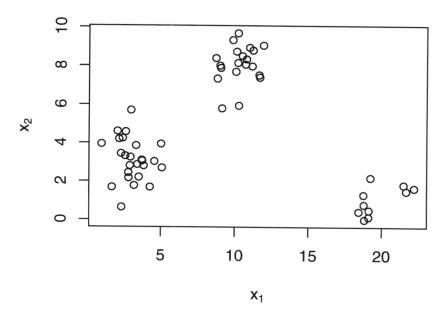

Fig. 6.1. Bivariate data showing the presence of three clusters.

6.3 Agglomerative hierarchical clustering

This class of clustering methods produces a *hierarchical classification* of data. In a hierarchical classification, the data are not partitioned into a particular number of classes or groups at a single step. Instead the classification consists of a series of partitions that may run from a single "cluster" containing all individuals to n clusters, each containing a single individual. Agglomerative hierarchical clustering techniques produce partitions by a series of successive fusions of the n individuals into groups. With such methods, fusions, once made, are irreversible, so that when an agglomerative algorithm has placed two individuals in the same group they cannot subsequently appear in different groups. Since all agglomerative hierarchical techniques ultimately reduce the data to a single cluster containing all the individuals, the investigator seeking the solution with the best-fitting number of clusters will need to decide which division to choose. The problem of deciding on the "correct" number of clusters will be taken up later.

An agglomerative hierarchical clustering procedure produces a series of partitions of the data, $P_n, P_{n-1}, \ldots, P_1$. The first, P_n, consists of n single-member clusters, and the last, P_1, consists of a single group containing all n individuals. The basic operation of all methods is similar:

(START) Clusters $C_1, C_2, \ldots, C_n$ each containing a single individual.
(1) Find the nearest pair of distinct clusters, say C_i and C_j, merge C_i and C_j, delete C_j, and decrease the number of clusters by one.
(2) If the number of clusters equals one, then stop; otherwise return to 1.

But before the process can begin, an inter-individual *distance matrix* or *similarity matrix* needs to be calculated. There are many ways to calculate distances or similarities between pairs of individuals, but here we only deal with a commonly used distance measure, Euclidean distance, which was defined in Chapter 1 but as a reminder is calculated as

$$d_{ij} = \sqrt{\sum_{k=1}^{q}(x_{ik} - x_{jk})^2},$$

where d_{ij} is the Euclidean distance between individual i with variable values $x_{i1}, x_{i2}, \ldots, x_{iq}$ and individual j with variable values $x_{j1}, x_{j2}, \ldots, x_{jq}$. (Details of other possible distance measures and similarity measures are given in Everitt et al. 2011). The Euclidean distances between each pair of individuals can be arranged in a matrix that is symmetric because $d_{ij} = d_{ji}$ and has zeros on the main diagonal. Such a matrix is the starting point of many clustering examples, although the calculation of Euclidean distances from the raw data may not be sensible when the variables are on very different scales. In such cases, the variables can be standardised in the usual way before calculating the distance matrix, although this can be unsatisfactory in some cases (see Everitt et al. 2011, for details).

Given an inter-individual distance matrix, the hierarchical clustering can begin, and at each stage in the process the methods fuse individuals or groups of individuals formed earlier that are closest (or most similar). So as groups are formed, the distance between an individual and a group containing several individuals and the distance between two groups of individuals will need to be calculated. How such distances are defined leads to a variety of different techniques. Two simple inter-group measures are

$$d_{AB} = \min_{\substack{i \in A \\ i \in B}} (d_{ij}),$$

$$d_{AB} = \max_{\substack{i \in A \\ i \in B}} (d_{ij}),$$

where d_{AB} is the distance between two clusters A and B, and d_{ij} is the distance between individuals i and j found from the initial inter-individual distance matrix.

The first inter-group distance measure above is the basis of *single linkage* clustering, the second that of *complete linkage* clustering. Both these techniques have the desirable property that they are invariant under monotone transformations of the original inter-individual distances; i.e., they only depend on the ranking on these distances, not their actual values.

A further possibility for measuring inter-cluster distance or dissimilarity is

$$d_{AB} = \frac{1}{n_A n_B} \sum_{i \in A} \sum_{i \in B} d_{ij},$$

where n_A and n_B are the numbers of individuals in clusters A and B. This measure is the basis of a commonly used procedure known as *group average* clustering. All three inter-group measures described above are illustrated in Figure 6.2.

Hierarchical classifications may be represented by a two-dimensional diagram known as a *dendrogram*, which illustrates the fusions made at each stage of the analysis. An example of such a diagram is given in Figure 6.3. The structure of Figure 6.3 resembles an *evolutionary tree*, a concept introduced by Darwin under the term "Tree of Life" in his book *On the Origin of Species by Natural Selection* in 1859, and it is in biological applications that hierarchical classifications are most relevant and most justified (although this type of clustering has also been used in many other areas).

As a first example of the application of the three clustering methods, single linkage, complete linkage, and group average, each will be applied to the chest, waist, and hip measurements of 20 individuals given in Chapter 1, Table 1.2. First Euclidean distances are calculated on the unstandardised measurements using the following R code:

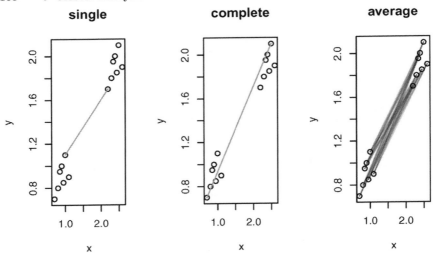

Fig. 6.2. Inter-cluster distance measures.

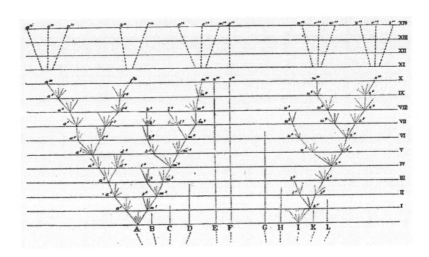

Fig. 6.3. Darwin's Tree of Life.

```
R> (dm <- dist(measure[, c("chest", "waist", "hips")]))
```

	1	2	3	4	5	6	7	8	9	10
2	6.16									
3	5.66	2.45								
4	7.87	2.45	4.69							
5	4.24	5.10	3.16	7.48						
6	11.00	6.08	5.74	7.14	7.68					
7	12.04	5.92	7.00	5.00	10.05	5.10				
8	8.94	3.74	4.00	3.74	7.07	5.74	4.12			
9	7.81	3.61	2.24	5.39	4.58	3.74	5.83	3.61		
10	10.10	4.47	4.69	5.10	7.35	2.24	3.32	3.74	3.00	
11	7.00	8.31	6.40	9.85	5.74	11.05	12.08	8.06	7.48	10.25
12	7.35	7.07	5.48	8.25	6.00	9.95	10.25	6.16	6.40	8.83
13	7.81	8.54	7.28	9.43	7.55	12.08	11.92	7.81	8.49	10.82
14	8.31	11.18	9.64	12.45	8.66	14.70	15.30	11.18	11.05	13.75
15	7.48	6.16	4.90	7.07	6.16	9.22	9.00	4.90	5.74	7.87
16	7.07	6.00	4.24	7.35	5.10	8.54	9.11	5.10	5.00	7.48
17	7.81	7.68	6.71	8.31	7.55	11.40	10.77	6.71	7.87	9.95
18	6.71	6.08	4.58	7.28	5.39	9.27	9.49	5.39	5.66	8.06
19	9.17	5.10	4.47	5.48	7.07	6.71	5.74	2.00	4.12	5.10
20	7.68	9.43	7.68	10.82	7.00	12.41	13.19	9.11	8.83	11.53

	11	12	13	14	15	16	17	18	19
2									
3									
4									
5									
6									
7									
8									
9									
10									
11									
12	2.24								
13	2.83	2.24							
14	3.74	5.20	3.74						
15	3.61	1.41	3.00	6.40					
16	3.00	1.41	3.61	6.40	1.41				
17	3.74	2.24	1.41	5.10	2.24	3.32			
18	2.83	1.00	2.83	5.83	1.00	1.00	2.45		
19	6.71	4.69	6.40	9.85	3.46	3.74	5.39	4.12	
20	1.41	3.00	2.45	2.45	4.36	4.12	3.74	3.74	7.68

Application of each of the three clustering methods described earlier to the distance matrix and a plot of the corresponding dendrogram are achieved using the hclust() function:

```
R> plot(cs <- hclust(dm, method = "single"))
R> plot(cc <- hclust(dm, method = "complete"))
R> plot(ca <- hclust(dm, method = "average"))
```

The resulting plots (for single, complete, and average linkage) are given in the upper part of Figure 6.4.

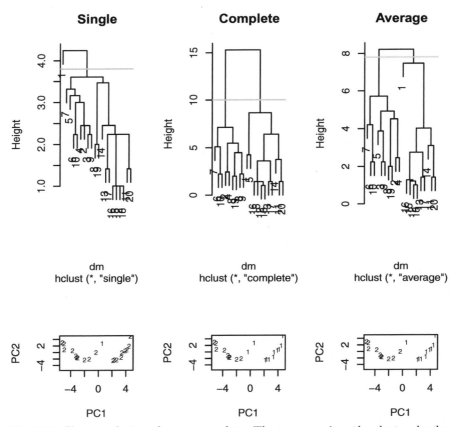

Fig. 6.4. Cluster solutions for measure data. The top row gives the cluster dendrograms along with the cutoff used to derive the classes presented (in the space of the first two principal components) in the bottom row.

We now need to consider how we select specific partitions of the data (i.e., a solution with a particular number of groups) from these dendrograms. The answer is that we "cut" the dendrogram at some height and this will give a partition with a particular number of groups. How do we choose where to cut or, in other words, how do we decide on a particular number of groups that is, in some sense, optimal for the data? This is a more difficult question to answer.

One informal approach is to examine the sizes of the changes in height in the dendrogram and take a "large" change to indicate the appropriate number of clusters for the data. (More formal approaches are described in Everitt et al. 2011) Even using this informal approach on the dendrograms in Figure 6.4, it is not easy to decide where to "cut".

So instead, because we know that these data consist of measurements on ten men and ten women, we will look at the two-group solutions from each method that are obtained by cutting the dendrograms at suitable heights. We can display and compare the three solutions graphically by plotting the first two principal component scores of the data, labelling the points to identify the cluster solution of one of the methods by using the following code:

```
R> body_pc <- princomp(dm, cor = TRUE)
R> xlim <- range(body_pc$scores[,1])
R> plot(body_pc$scores[,1:2], type = "n",
+        xlim = xlim, ylim = xlim)
R> lab <- cutree(cs, h = 3.8)
R> text(body_pc$scores[,1:2], labels = lab, cex = 0.6)
```

The resulting plots are shown in the lower part of Figure 6.4. The plots of dendrograms and principal components scatterplots are combined into a single diagram using the layout() function (see the chapter demo for the complete R code). The plot associated with the single linkage solution immediately demonstrates one of the problems with using this method in practise, and that is a phenomenon known as *chaining*, which refers to the tendency to incorporate intermediate points between clusters into an existing cluster rather than initiating a new one. As a result, single linkage solutions often contain long "straggly" clusters that do not give a useful description of the data. The two-group solutions from complete linkage and average linkage, also shown in Figure 6.4, are similar and in essence place the men (observations 1 to 10) together in one cluster and the women (observations 11 to 20) in the other.

6.3.1 Clustering jet fighters

The data shown in Table 6.1 as originally given in Stanley and Miller (1979) and also in Hand et al. (1994) are the values of six variables for 22 US fighter aircraft. The variables are as follows:

FFD: first flight date, in months after January 1940;
SPR: specific power, proportional to power per unit weight;
RGF: flight range factor;
PLF: payload as a fraction of gross weight of aircraft;
SLF: sustained load factor;
CAR: a binary variable that takes the value 1 if the aircraft can land on a carrier and 0 otherwise.

Table 6.1: jet data. Jet fighters data.

FFD	SPR	RGF	PLF	SLF	CAR
82	1.468	3.30	0.166	0.10	no
89	1.605	3.64	0.154	0.10	no
101	2.168	4.87	0.177	2.90	yes
107	2.054	4.72	0.275	1.10	no
115	2.467	4.11	0.298	1.00	yes
122	1.294	3.75	0.150	0.90	no
127	2.183	3.97	0.000	2.40	yes
137	2.426	4.65	0.117	1.80	no
147	2.607	3.84	0.155	2.30	no
166	4.567	4.92	0.138	3.20	yes
174	4.588	3.82	0.249	3.50	no
175	3.618	4.32	0.143	2.80	no
177	5.855	4.53	0.172	2.50	yes
184	2.898	4.48	0.178	3.00	no
187	3.880	5.39	0.101	3.00	yes
189	0.455	4.99	0.008	2.64	no
194	8.088	4.50	0.251	2.70	yes
197	6.502	5.20	0.366	2.90	yes
201	6.081	5.65	0.106	2.90	yes
204	7.105	5.40	0.089	3.20	yes
255	8.548	4.20	0.222	2.90	no
328	6.321	6.45	0.187	2.00	yes

We shall apply complete linkage to the data but using only variables two to five. And given that the variables are on very different scales, we will standardise them to unit variance before clustering. The required R code for standardisation and clustering is as follows:

```
R> X <- scale(jet[, c("SPR", "RGF", "PLF", "SLF")],
+           center = FALSE, scale = TRUE)
R> dj <- dist(X)
R> plot(cc <- hclust(dj), main = "Jets clustering")
R> cc

Call:
hclust(d = dj)


Cluster method   : complete
Distance         : euclidean
Number of objects: 22
```

The resulting dendrogram in Figure 6.5 strongly suggests the presence of two groups of fighters. In Figure 6.6, the data are plotted in the space of the first two principal components of the correlation matrix of the relevant variables (SPR to SLF). And in Figure 6.6 the points are labelled by cluster number for the two-group solution and the colours used are the values of the CAR variable. The two-group solution largely corresponds to planes that can and cannot land on a carrier.

```
Call:
hclust(d = dj)

Cluster method    : complete
Distance          : euclidean
Number of objects: 22
```

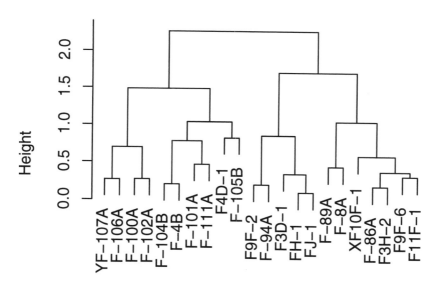

Fig. 6.5. Hierarchical clustering (complete linkage) of jet data.

Again, we cut the dendrogram in such a way that two clusters remain and plot the corresponding classes in the space of the first two principal components; see Figure 6.6.

```
R> pr <- prcomp(dj)$x[, 1:2]
R> plot(pr, pch = (1:2)[cutree(cc, k = 2)],
+       col = c("black", "darkgrey")[jet$CAR],
+       xlim = range(pr) * c(1, 1.5))
R> legend("topright", col = c("black", "black",
+                              "darkgrey", "darkgrey"),
+         legend = c("1 / no", "2 / no", "1 / yes", "2 / yes"),
+         pch = c(1:2, 1:2), title = "Cluster / CAR", bty = "n")
```

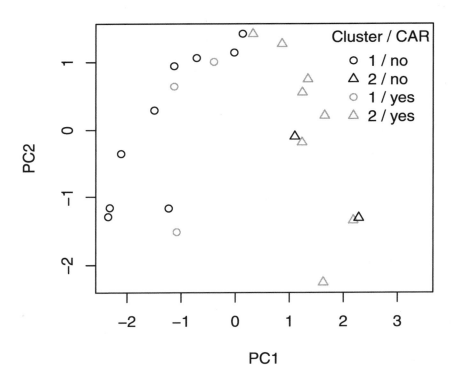

Fig. 6.6. Hierarchical clustering (complete linkage) of jet data plotted in PCA space.

6.4 *K*-means clustering

The k-means clustering technique seeks to partition the n individuals in a set of multivariate data into k groups or clusters, $(G_1, G_2, \ldots, G_k)$, where G_i denotes the set of n_i individuals in the ith group, and k is given (or a possible range is specified by the researcher–the problem of choosing the "true" value of k will be taken up later) by minimising some numerical criterion, low values of which are considered indicative of a "good" solution. The most commonly used implementation of k-means clustering is one that tries to find the partition of the n individuals into k groups that minimises the *within-group sum of squares* (WGSS) over all variables; explicitly, this criterion is

$$\text{WGSS} = \sum_{j=1}^{q} \sum_{l=1}^{k} \sum_{i \in G_l} (x_{ij} - \overline{x}_j^{(l)})^2,$$

where $\overline{x}_j^{(l)} = \frac{1}{n_i} \sum_{i \in G_l} x_{ij}$ is the mean of the individuals in group G_l on variable j.

The problem then appears relatively simple; namely, consider every possible partition of the n individuals into k groups, and select the one with the lowest within-group sum of squares. Unfortunately, the problem in practise is not so straightforward. The numbers involved are so vast that complete enumeration of *every* possible partition remains impossible even with the fastest computer. The scale of the problem immediately becomes clear by looking at the numbers in Table 6.2.

Table 6.2: Number of possible partitions depending on the sample size n and number of clusters k.

n	k	Number of possible partitions
15	3	$2,375,101$
20	4	$45,232,115,901$
25	8	$690,223,721,118,368,580$
100	5	10^{68}

The impracticability of examining every possible partition has led to the development of algorithms designed to search for the minimum values of the clustering criterion by rearranging existing partitions and keeping the new one only if it provides an improvement. Such algorithms do not, of course, guarantee finding the global minimum of the criterion. The essential steps in these algorithms are as follows:

1. Find some initial partition of the individuals into the required number of groups. (Such an initial partition could be provided by a solution from one of the hierarchical clustering techniques described in the previous section.)
2. Calculate the change in the clustering criterion produced by "moving" each individual from its own cluster to another cluster.
3. Make the change that leads to the greatest improvement in the value of the clustering criterion.
4. Repeat steps (2) and (3) until no move of an individual causes the clustering criterion to improve.

(For a more detailed account of the typical k-means algorithm see Steinley 2008)

The k-means approach to clustering using the minimisation of the within-group sum of squares over all the variables is widely used but suffers from the two problems of (1) not being scale-invariant (i.e., different solutions may result from clustering the raw data and the data standardised in some way) and (2) of imposing a "spherical" structure on the data; i.e., it will find clusters shaped like hyper-footballs even if the "true" clusters in the data are of some other shape (see Everitt et al. 2011, for some examples of the latter phenomenon). Nevertheless, the k-means method remains very popular. With k-means clustering, the investigator can choose to partition the data into a specified number of groups. In practice, solutions for a range of values for the number of groups are found and in some way the optimal or "true" number of groups for the data must be chosen. Several suggestions have been made as to how to answer the number of groups question, but none is completely satisfactory. The method we shall use in the forthcoming example is to plot the within-groups sum of squares associated with the k-means solution for each number of groups. As the number of groups increases the sum of squares will necessarily decrease, but an obvious "elbow" in the plot may be indicative of the most useful solution for the investigator to look at in detail. (Compare this with the scree plot described in Chapter 3.)

6.4.1 Clustering the states of the USA on the basis of their crime rate profiles

The *Statistical Abstract of the USA* (Anonymous 1988, Table 265) gives rates of different types of crime per 100,000 residents of the 50 states of the USA plus the District of Columbia for the year 1986. The data are given in Table 6.3.

Table 6.3: `crime` data. Crime data.

	Murder	Rape	Robbery	Assault	Burglary	Theft	Vehicle
ME	2.0	14.8	28	102	803	2347	164
NH	2.2	21.5	24	92	755	2208	228
VT	2.0	21.8	22	103	949	2697	181
MA	3.6	29.7	193	331	1071	2189	906

Table 6.3: crime data (continued).

	Murder	Rape	Robbery	Assault	Burglary	Theft	Vehicle
RI	3.5	21.4	119	192	1294	2568	705
CT	4.6	23.8	192	205	1198	2758	447
NY	10.7	30.5	514	431	1221	2924	637
NJ	5.2	33.2	269	265	1071	2822	776
PA	5.5	25.1	152	176	735	1654	354
OH	5.5	38.6	142	235	988	2574	376
IN	6.0	25.9	90	186	887	2333	328
IL	8.9	32.4	325	434	1180	2938	628
MI	11.3	67.4	301	424	1509	3378	800
WI	3.1	20.1	73	162	783	2802	254
MN	2.5	31.8	102	148	1004	2785	288
IA	1.8	12.5	42	179	956	2801	158
MO	9.2	29.2	170	370	1136	2500	439
ND	1.0	11.6	7	32	385	2049	120
SD	4.0	17.7	16	87	554	1939	99
NE	3.1	24.6	51	184	748	2677	168
KS	4.4	32.9	80	252	1188	3008	258
DE	4.9	56.9	124	241	1042	3090	272
MD	9.0	43.6	304	476	1296	2978	545
DC	31.0	52.4	754	668	1728	4131	975
VA	7.1	26.5	106	167	813	2522	219
WV	5.9	18.9	41	99	625	1358	169
NC	8.1	26.4	88	354	1225	2423	208
SC	8.6	41.3	99	525	1340	2846	277
GA	11.2	43.9	214	319	1453	2984	430
FL	11.7	52.7	367	605	2221	4373	598
KY	6.7	23.1	83	222	824	1740	193
TN	10.4	47.0	208	274	1325	2126	544
AL	10.1	28.4	112	408	1159	2304	267
MS	11.2	25.8	65	172	1076	1845	150
AR	8.1	28.9	80	278	1030	2305	195
LA	12.8	40.1	224	482	1461	3417	442
OK	8.1	36.4	107	285	1787	3142	649
TX	13.5	51.6	240	354	2049	3987	714
MT	2.9	17.3	20	118	783	3314	215
ID	3.2	20.0	21	178	1003	2800	181
WY	5.3	21.9	22	243	817	3078	169
CO	7.0	42.3	145	329	1792	4231	486
NM	11.5	46.9	130	538	1845	3712	343
AZ	9.3	43.0	169	437	1908	4337	419
UT	3.2	25.3	59	180	915	4074	223
NV	12.6	64.9	287	354	1604	3489	478

Table 6.3: `crime` data (continued).

	Murder	Rape	Robbery	Assault	Burglary	Theft	Vehicle
WA	5.0	53.4	135	244	1861	4267	315
OR	6.6	51.1	206	286	1967	4163	402
CA	11.3	44.9	343	521	1696	3384	762
AK	8.6	72.7	88	401	1162	3910	604
HI	4.8	31.0	106	103	1339	3759	328

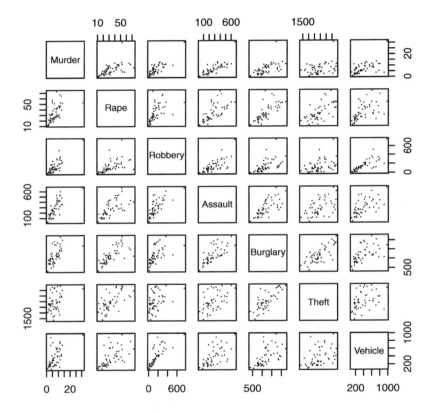

Fig. 6.7. Scatterplot matrix of crime data.

To begin, let's look at the scatterplot matrix of the data shown in Figure 6.7. The plot suggests that at least one of the cities is considerably different from the others in its murder rate at least. The city is easily identified using

```
R> subset(crime, Murder > 15)
```

	Murder	Rape	Robbery	Assault	Burglary	Theft	Vehicle
DC	31	52.4	754	668	1728	4131	975

i.e., the murder rate is very high in the District of Columbia. In order to check if the other crime rates are also higher in DC, we label the corresponding points in the scatterplot matrix in Figure 6.8. Clearly, DC is rather extreme in most crimes (the clear message is don't live in DC).

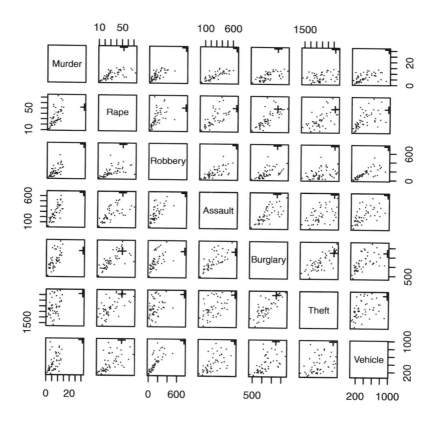

Fig. 6.8. Scatterplot matrix of crime data with DC observation labelled using a plus sign.

We will now apply *k*-means clustering to the crime rate data *after* removing the outlier, DC. If we first calculate the variances of the crime rates for the different types of crimes we find the following:

```
R> sapply(crime, var)
```

Murder	Rape	Robbery	Assault	Burglary	Theft	Vehicle
23.2	212.3	18993.4	22004.3	177912.8	582812.8	50007.4

The variances are very different, and using k-means on the raw data would not be sensible; we must standardise the data in some way, and here we standardise each variable by its range. After such standardisation, the variances become

```R
R> rge <- sapply(crime, function(x) diff(range(x)))
R> crime_s <- sweep(crime, 2, rge, FUN = "/")
R> sapply(crime_s, var)
```

Murder	Rape	Robbery	Assault	Burglary	Theft	Vehicle
0.02578	0.05687	0.03404	0.05440	0.05278	0.06411	0.06517

The variances of the standardised data are very similar, and we can now progress with clustering the data. First we plot the within-groups sum of squares for one- to six-group solutions to see if we can get any indication of the number of groups. The plot is shown in Figure 6.9. The only "elbow" in the plot occurs for two groups, and so we will now look at the two-group solution. The group means for two groups are computed by

```R
R> kmeans(crime_s, centers = 2)$centers * rge
```

	Murder	Rape	Robbery	Assault	Burglary	Theft	Vehicle
1	4.893	305.1	189.6	259.70	31.0	540.5	873.0
2	21.098	483.3	1031.4	19.26	638.9	2096.1	578.6

A plot of the two-group solution in the space of the first two principal components of the correlation matrix of the data is shown in Figure 6.10. The two groups are created essentially on the basis of the first principal component score, which is a weighted average of the crime rates. Perhaps all the cluster analysis is doing here is dividing into two parts a homogenous set of data. This is always a possibility, as is discussed in some detail in Everitt et al. (2011).

6.4.2 Clustering Romano-British pottery

The second application of k-means clustering will be to the data on Romano-British pottery given in Chapter 1. We begin by computing the Euclidean distance matrix for the standardised measurements of the 45 pots. The resulting 45×45 matrix can be inspected graphically by using an *image plot*, here obtained with the function levelplot available in the package **lattice** (Sarkar 2010, 2008). Such a plot associates each cell of the dissimilarity matrix with a colour or a grey value. We choose a very dark grey for cells with distance zero (i.e., the diagonal elements of the dissimilarity matrix) and pale values for cells with greater Euclidean distance. Figure 6.11 leads to the impression that there are at least three distinct groups with small inter-cluster differences (the dark rectangles), whereas much larger distances can be observed for all other cells.

```
R> n <- nrow(crime_s)
R> wss <- rep(0, 6)
R> wss[1] <- (n - 1) * sum(sapply(crime_s, var))
R> for (i in 2:6)
+        wss[i] <- sum(kmeans(crime_s,
+                            centers = i)$withinss)
R> plot(1:6, wss, type = "b", xlab = "Number of groups",
+        ylab = "Within groups sum of squares")
```

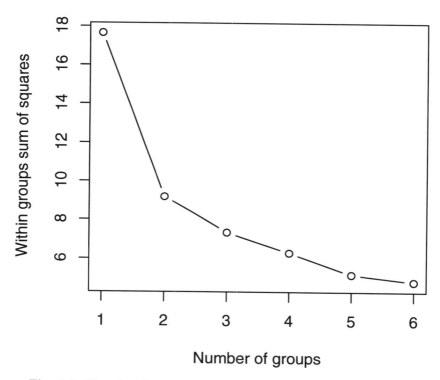

Fig. 6.9. Plot of within-groups sum of squares against number of clusters.

We plot the within-groups sum of squares for one to six group *k*-means solutions to see if we can get any indication of the number of groups (see Figure 6.12). Again, the plot leads to the relatively clear conclusion that the data contain three clusters.

Our interest is now in a comparison of the kiln sites at which the pottery was found.

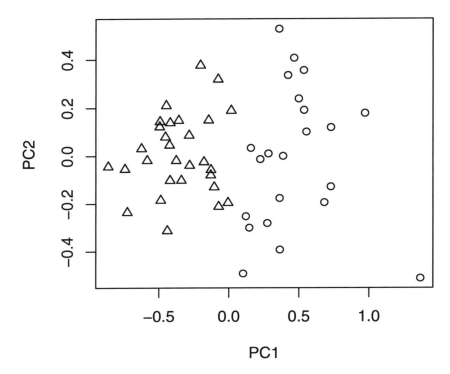

Fig. 6.10. Plot of k-means two-group solution for the standardised crime rate data.

```
R> set.seed(29)
R> pottery_cluster <- kmeans(pots, centers = 3)$cluster
R> xtabs(~ pottery_cluster + kiln, data = pottery)

                kiln
pottery_cluster  1  2  3  4  5
              1 21  0  0  0  0
              2  0 12  2  0  0
              3  0  0  0  5  5
```

The contingency table shows that cluster 1 contains all pots found at kiln site number one, cluster 2 contains all pots from kiln sites numbers two and three, and cluster three collects the ten pots from kiln sites four and five. In fact, the five kiln sites are from three different regions: region 1 contains just kiln one, region 2 contains kilns two and three, and region 3 contains kilns four

```
R> pottery_dist <- dist(pots <- scale(pottery[, colnames(pottery) != "kiln"],
+                                 center = FALSE))
R> library("lattice")
R> levelplot(as.matrix(pottery_dist), xlab = "Pot Number",
+            ylab = "Pot Number")
```

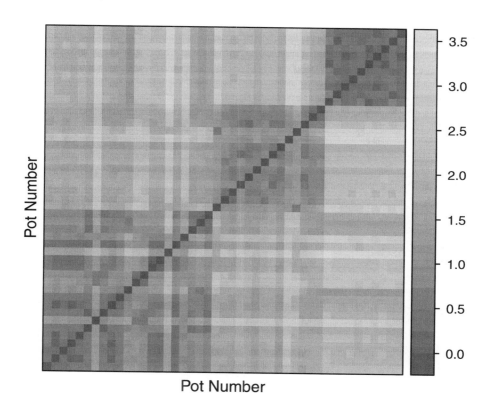

Fig. 6.11. Image plot of the dissimilarity matrix of the `pottery` data.

and five. So the clusters found actually correspond to pots from three different regions.

6.5 Model-based clustering

The agglomerative hierarchical and k-means clustering methods described in the previous two sections are based largely on heuristic but intuitively reasonable procedures. But they are not based on formal models for cluster structure in the data, making problems such as deciding between methods, estimating

```
R> n <- nrow(pots)
R> wss <- rep(0, 6)
R> wss[1] <- (n - 1) * sum(sapply(pots, var))
R> for (i in 2:6)
+        wss[i] <- sum(kmeans(pots,
+                             centers = i)$withinss)
R> plot(1:6, wss, type = "b", xlab = "Number of groups",
+        ylab = "Within groups sum of squares")
```

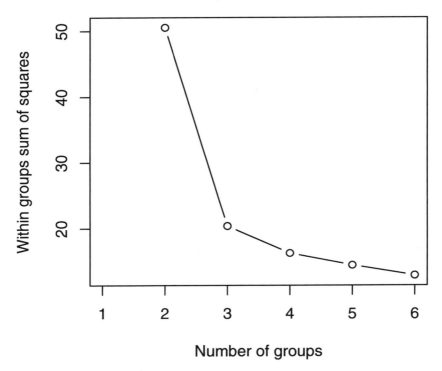

Fig. 6.12. Plot of within-groups sum of squares against number of clusters.

the number of clusters, etc, particularly difficult. And, of course, without a reasonable model, formal inference is precluded. In practise, these may not be insurmountable objections to the use of either the agglomerative methods or k-means clustering because cluster analysis is most often used as an "exploratory" tool for data analysis. But if an acceptable model for cluster structure could be found, then the cluster analysis based on the model might give more persuasive solutions (more persuasive to statisticians at least). In

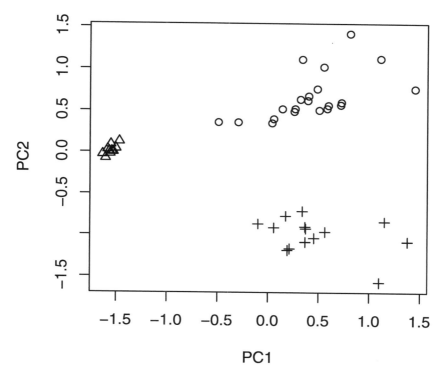

Fig. 6.13. Plot of the k-means three-group solution for the pottery data displayed in the space of the first two principal components of the correlation matrix of the data.

this section, we describe an approach to clustering that postulates a formal statistical model for the population from which the data are sampled, a model that assumes that this population consists of a number of subpopulations (the "clusters"), each having variables with a different multivariate probability density function, resulting in what is known as a *finite mixture density* for the population as a whole. By using finite mixture densities as models for cluster analysis, the clustering problem becomes that of estimating the parameters of the assumed mixture and then using the estimated parameters to calculate the posterior probabilities of cluster membership. And determining the number of clusters reduces to a model selection problem for which objective procedures exist.

Finite mixture densities often provide a sensible statistical *model* for the clustering process, and cluster analyses based on finite mixture models are also known as *model-based* clustering methods; see Banfield and Raftery (1993). Finite mixture models have been increasingly used in recent years to cluster data in a variety of disciplines, including behavioural, medical, genetic, computer, environmental sciences, and robotics and engineering; see, for example, Everitt and Bullmore (1999), Bouguila and Amayri (2009), Branchaud, Cham, Nenadic, Andersen, and Burdick (2010), Dai, Erkkila, Yli-Harja, and Lahdesmaki (2009), Dunson (2009), Ganesalingam, Stahl, Wijesekera, Galtrey, Shaw, Leigh, and Al-Chalabi (2009), Marin, Mengersen, and Roberts (2005), Meghani, Lee, Hanlon, and Bruner (2009), Pledger and Phillpot (2008), and van Hattum and Hoijtink (2009).

Finite mixture modelling can be seen as a form of *latent variable analysis* (see, for example, Skrondal and Rabe-Hesketh 2004), with "subpopulation" being a latent categorical variable and the latent classes being described by the different components of the mixture density; consequently, cluster analysis based on such models is also often referred to as *latent class cluster analysis*.

6.5.1 Finite mixture densities

Finite mixture densities are described in detail in Everitt and Hand (1981), Titterington, Smith, and Makov (1985), McLachlan and Basford (1988), McLachlan and Peel (2000), and Frühwirth-Schnatter (2006); they are a family of probability density functions of the form

$$f(\mathbf{x}; \mathbf{p}, \boldsymbol{\theta}) = \sum_{j=1}^{c} p_j g_j(\mathbf{x}; \boldsymbol{\theta}_j), \tag{6.1}$$

where $\mathbf{x}$ is a p-dimensional random variable, $\mathbf{p}^{\top} = (p_1, p_2, \ldots, p_{c-1})$, and $\boldsymbol{\theta}^{\top} = (\boldsymbol{\theta}_1^{\top}, \boldsymbol{\theta}_2^{\top}, \ldots, \boldsymbol{\theta}_c^{\top})$, with the p_j being known as mixing proportions and the g_j, $j = 1, \ldots, c$, being the component densities, with density g_j being parameterised by $\boldsymbol{\theta}_j$. The mixing proportions are non-negative and are such that $\sum_{j=1}^{c} p_j = 1$. The number of components forming the mixture (i.e., the postulated number of clusters) is c.

Finite mixtures provide suitable models for cluster analysis if we assume that each group of observations in a data set suspected to contain clusters comes from a population with a different probability distribution. The latter may belong to the same family but differ in the values they have for the parameters of the distribution; it is such an example that we consider in the next section, where the components of the mixture are multivariate normal with different mean vectors and possibly different covariance matrices.

Having estimated the parameters of the assumed mixture density, observations can be associated with particular clusters on the basis of the maximum value of the estimated posterior probability

$$\hat{\mathsf{P}}(\text{cluster } j | \mathbf{x}_i) = \frac{\hat{p}_j g_j(\mathbf{x}_i; \hat{\boldsymbol{\theta}}_j)}{f(\mathbf{x}_i; \hat{\mathbf{p}}, \hat{\boldsymbol{\theta}})}, j = 1, \ldots, c. \tag{6.2}$$

6.5.2 Maximum likelihood estimation in a finite mixture density with multivariate normal components

Given a sample of observations $\mathbf{x}_1, \mathbf{x}_2, \ldots, \mathbf{x}_n$, from the mixture density given in Equation (6.1) the log-likelihood function, l, is

$$l(\mathbf{p}, \boldsymbol{\theta}) = \sum_{i=1}^{n} \ln f(\mathbf{x}_i; \mathbf{p}, \boldsymbol{\theta}). \tag{6.3}$$

Estimates of the parameters in the density would usually be obtained as a solution of the likelihood equations

$$\frac{\partial l(\boldsymbol{\varphi})}{\partial(\boldsymbol{\varphi})} = 0, \tag{6.4}$$

where $\boldsymbol{\varphi}^{\top} = (\mathbf{p}^{\top}, \boldsymbol{\theta}^{\top})$. In the case of finite mixture densities, the likelihood function is too complicated to employ the usual methods for its maximisation; for example, an iterative Newton–Raphson method that approximates the gradient vector of the log-likelihood function $l(\boldsymbol{\varphi})$ by a linear Taylor series expansion (see Everitt (1984)).

Consequently, the required maximum likelihood estimates of the parameters in a finite mixture model have to be computed in some other way. In the case of a mixture in which the jth component density is multivariate normal with mean vector $\boldsymbol{\mu}_j$ and covariance matrix $\boldsymbol{\Sigma}_j$, it can be shown (see Everitt and Hand 1981, for details) that the application of maximum likelihood results in the series of equations

$$\hat{p}_j = \frac{1}{n} \sum_{i=1}^{n} \hat{P}(j|\mathbf{x}_i), \tag{6.5}$$

$$\hat{\boldsymbol{\mu}}_j = \frac{1}{n\hat{p}_j} \sum_{i=1}^{n} \mathbf{x}_i \hat{P}(j|\mathbf{x}_i), \tag{6.6}$$

$$\hat{\boldsymbol{\Sigma}}_j = \frac{1}{n} \sum_{i=1}^{n} (\mathbf{x}_i - \boldsymbol{\mu}_j)(\mathbf{x}_i - \boldsymbol{\mu}_j)^{\top} \hat{P}(j|\mathbf{x}_i), \tag{6.7}$$

where the $\hat{P}(j|\mathbf{x}_i)$s are the estimated posterior probabilities given in equation (6.2).

Hasselblad (1966, 1969), Wolfe (1970), and Day (1969) all suggest an iterative scheme for solving the likelihood equations given above that involves finding initial estimates of the posterior probabilities given initial estimates of the parameters of the mixture and then evaluating the right-hand sides of Equations 6.5 to 6.7 to give revised values for the parameters. From these, new estimates of the posterior probabilities are derived, and the procedure is repeated until some suitable convergence criterion is satisfied. There are potential problems with this process unless the component covariance matrices are constrained in some way; for example, it they are all assumed to be the same–again see Everitt and Hand (1981) for details.

This procedure is a particular example of the iterative *expectation maximisation* (EM) algorithm described by Dempster, Laird, and Rubin (1977) in the context of likelihood estimation for incomplete data problems. In estimating parameters in a mixture, it is the "labels" of the component density from which an observation arises that are missing. As an alternative to the EM algorithm, Bayesian estimation methods using the Gibbs sampler or other Monte Carlo Markov Chain (MCMC) methods are becoming increasingly popular–see Marin et al. (2005) and McLachlan and Peel (2000).

Fraley and Raftery (2002, 2007) developed a series of finite mixture density models with multivariate normal component densities in which they allow some, but not all, of the features of the covariance matrix (*orientation, size,* and *shape*–discussed later) to vary between clusters while constraining others to be the same. These new criteria arise from considering the reparameterisation of the covariance matrix Σ_j in terms of its eigenvalue description

$$\Sigma_j = \mathbf{D}_j \Lambda_j \mathbf{D}_j^\top, \qquad (6.8)$$

where $\mathbf{D}_j$ is the matrix of eigenvectors and Λ_j is a diagonal matrix with the eigenvalues of Σ_j on the diagonal (this is simply the usual principal components transformation–see Chapter 3). The orientation of the principal components of Σ_j is determined by $\mathbf{D}_j$, whilst Λ_j specifies the size and shape of the density contours. Specifically, we can write $\Lambda_j = \lambda_j \mathbf{A}_j$, where λ_j is the largest eigenvalue of Σ_j and $\mathbf{A}_j = \mathrm{diag}(1, \alpha_2, \ldots, \alpha_p)$ contains the eigenvalue ratios after division by λ_j. Hence λ_j controls the size of the jth cluster and $\mathbf{A}_j$ its shape. (Note that the term "size" here refers to the volume occupied in space, not the number of objects in the cluster.) In two dimensions, the parameters would reflect, for each cluster, the correlation between the two variables, and the magnitudes of their standard deviations. More details are given in Banfield and Raftery (1993) and Celeux and Govaert (1995), but Table 6.4 gives a series of models corresponding to various constraints imposed on the covariance matrix. The models make up what Fraley and Raftery (2003, 2007) term the "MCLUST" family of mixture models. The mixture likelihood approach based on the EM algorithm for parameter estimation is implemented in the Mclust() function in the R package **mclust** and fits the models in the MCLUST family described in Table 6.4.

Model selection is a combination of choosing the appropriate clustering model for the population from which the n observations have been taken (i.e., are all clusters spherical, all elliptical, all different shapes or somewhere in between?) *and* the optimal number of clusters. A Bayesian approach is used (see Fraley and Raftery 2002), applying what is known as the *Bayesian Information Criterion* (BIC). The result is a cluster solution that "fits" the observed data as well as possible, and this can include a solution that has only one "cluster" implying that cluster analysis is not really a useful technique for the data.

Table 6.4: **mclust** family of mixture models. Model names describe model restrictions of volume λ_j, shape $\mathbf{A}_j$, and orientation $\mathbf{D}_j$, $V=$ variable, parameter unconstrained, $E=$ equal, parameter constrained, $I =$ matrix constrained to identity matrix.

Abbreviation	Model
EII	spherical, equal volume
VII	spherical, unequal volume
EEI	diagonal, equal volume and shape
VEI	diagonal, varying volume, equal shape
EVI	diagonal, equal volume, varying shape
VVI	diagonal, varying volume and shape
EEE	ellipsoidal, equal volume, shape, and orientation
EEV	ellipsoidal, equal volume and equal shape
VEV	ellipsoidal, equal shape
VVV	ellipsoidal, varying volume, shape, and orientation

To illustrate the use of the finite mixture approach to cluster analysis, we will apply it to data that arise from a study of what gastroenterologists in Europe tell their cancer patients (Thomsen, Wulff, Martin, and Singer 1993). A questionnaire was sent to about 600 gastroenterologists in 27 European countries (the study took place before the recent changes in the political map of the continent) asking what they would tell a patient with newly diagnosed cancer of the colon, and his or her spouse, about the diagnosis. The respondent gastroenterologists were asked to read a brief case history and then to answer six questions with a yes/no answer. The questions were as follows:

Q1: Would you tell this patient that he/she has cancer, if he/she asks no questions?

Q2: Would you tell the wife/husband that the patient has cancer (In the patient's absence)?

Q3: Would you tell the patient that he or she has a cancer, if he or she directly asks you to disclose the diagnosis. (During surgery the surgeon notices several small metastases in the liver.)

Q4: Would you tell the patient about the metastases (supposing the patient asks to be told the results of the operation)?

Q5: Would you tell the patient that the condition is incurable?

Q6: Would you tell the wife or husband that the operation revealed metastases?

The data are shown in a graphical form in Figure 6.14 (we are aware that using finite mixture clustering on this type of data is open to criticism–it may even be a statistical sin–but we hope that even critics will agree it provides an interesting example).

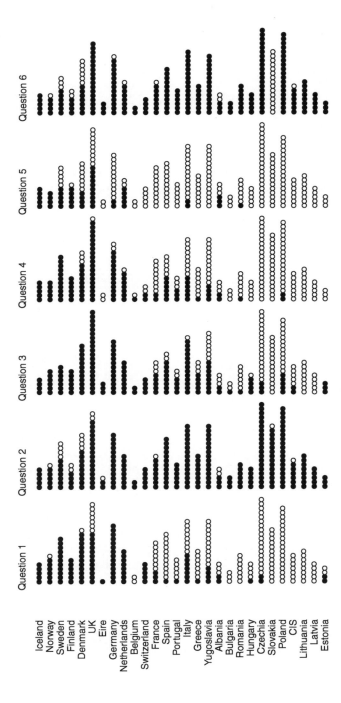

Fig. 6.14. Gastroenterologists questionnaire data. Dark circles indicate a 'yes', open circles a 'no'.

Applying the finite mixture approach to the proportions of 'yes' answers for each question for each country computed from these data using the R code utilizing functionality offered by package **mclust** (Fraley and Raftery 2010)

```
R> library("mclust")
```

by using mclust, invoked on its own or through another package, you accept the license agreement in the mclust LICENSE file and at http://www.stat.washington.edu/mclust/license.txt

```
R> (mc <- Mclust(thomsonprop))
```

```
best model: ellipsoidal, equal shape with 3 components
```

where `thomsonprob` is the matrix of proportions of "yes" answers to questions Q1–Q6 in the different countries (i.e., the proportion of filled dots in Figure 6.14) available from the **MVA** add-on package. We can first examine the resulting plot of BIC values shown in Figure 6.15. In this diagram, the plot symbols and abbreviations refer to different model assumptions about the shapes of clusters as given in Table 6.4.

The BIC criterion selects model VEV (ellipsoidal, equal shape) and three clusters as the optimal solution. The three-cluster solution is illustrated graphically in Figure 6.16. The first cluster consists of countries in which the large majority of respondents gave "yes" answers to questions $1, 2, 3, 4$, and 6 and about half also gave a "yes" answer to question 5. This cluster includes all the Scandinavian countries the UK, Iceland, Germany, the Netherlands, and Switzerland. In the second cluster, the majority of respondents answer "no" to questions $1, 4$, and 5 and "yes" to questions $2, 3$ and 6; in these countries it appears that the clinicians do not mind giving bad news to the spouses of patients but not to the patients themselves unless they are directly asked by the patient about hispr her condition. This cluster contains Catholic countries such as Spain, Portugal, and Italy. In cluster three, the large majority of respondents answer "no" to questions $1, 3, 4$, and 5 and again a large majority answer "yes" to questions 2 and 6. In these countries, very few clinicians appear to be willing to give the patient bad news even if asked directly by the patient about his or her condition.

6.6 Displaying clustering solutions graphically

Plotting cluster solutions in the space of the first few principal components as illustrated earlier in this chapter is often a useful way to display clustering solutions, but other methods of displaying clustering solutions graphically are also available. Leisch (2010), for example, describes several graphical displays that can be used to visualise cluster analysis solutions. The basis of a number of these graphics is the shadow value, $s(\mathbf{x})$, of each multivariate observation, $\mathbf{x}$, defined as

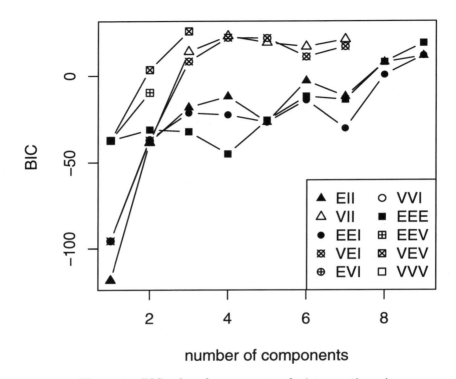

Fig. 6.15. BIC values for gastroenterologists questionnaire.

$$s(\mathbf{x}) = \frac{2d(\mathbf{x}, c(\mathbf{x}))}{d(\mathbf{x}, c(\mathbf{x})) + d(\mathbf{x}, \tilde{c}(\mathbf{x}))},$$

where $d(\mathbf{x}, c(\mathbf{x}))$ is the distance of the observation $\mathbf{x}$ from the centroid of its own cluster and $d(\mathbf{x}, \tilde{c}(\mathbf{x}))$ is the distance of $\mathbf{x}$ from the second closest cluster centroid. If $s(\mathbf{x})$ is close to zero, then the observation is close to its cluster centroid; if $s(\mathbf{x})$ is close to one, then the observation is almost equidistant from the two centroids (a similar approach is used in defining silhouette plots, see Chapter 5). The average shadow value of all observations where cluster i is closest and cluster j is second closest can be used as a simple measure of cluster similarity,

$$s_{ij} = \frac{1}{n_i} \sum_{\mathbf{x} \in A_{ij}} s(\mathbf{x}),$$

where n_i is the number of observations that are closest to the centroid of cluster i and A_{ij} is the set of observations for which the centroid of cluster i is

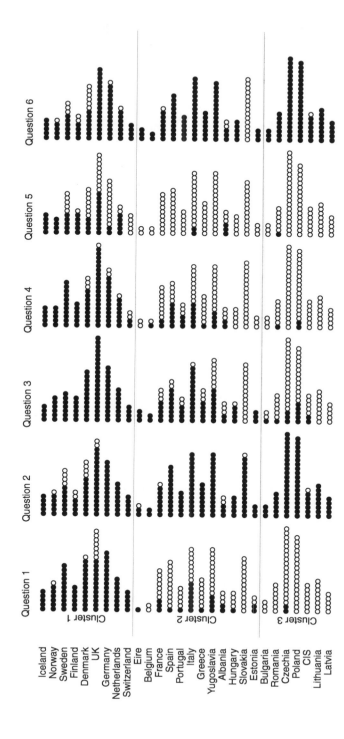

Fig. 6.16. Model-based clustering into three clusters for gastroenterologists questionnaire data. Dark circles indicate a 'yes', open circles a 'no'.

closest and the centroid of cluster j is second closest. The denominator of s_{ij} is taken to be n_i rather than n_{ij}, the number of observations in the set A_{ij}, to prevent inducing large cluster similarity when n_{ij} is small and the set of observations consists of poorly clustered points with large shadow values. For a cluster solution derived from bivariate data, a neighbourhood graph can be constructed using the scatterplot of the two variables, and where two cluster centroids are joined if there exists at least one observation for which these two are closest, and second closest with the thickness of the joining lines being made proportional to the average value of the corresponding s_{ij}. When there are more than two variables in the data set, the neighbourhood graph can be constructed on some suitable projection of the data into two dimensions; for example, the first two principal components of the data could be used. Such plots may help to establish which clusters are "real" and which are not, as we will try to illustrate with two examples.

The first example uses some two-dimensional data generated to contain three clusters. The neighbourhood graph for the k-means five-cluster solution from the application of k-means clustering is shown in Figure 6.17. The thicker lines joining the centroids of clusters 1 and 5 and clusters 2 and 4 strongly suggest that both pairs of clusters overlap to a considerable extent and are probably each divisions of a single cluster.

For the second example we return to the pottery data previously met in the chapter. From the k-means analysis, it is clear that these data contain three clusters; Figure 6.18 shows the neighbourhood plot for the k-means three-cluster solution in the space of the first two principal components of the data. The three clusters are clearly visible in this plot.

A further graphic for displaying clustering solutions is known as a stripes plot. This graphic is a simple but often effective way of visualising the distance of each point from its closest and second closest cluster centroids. For each cluster, $k = 1, \ldots, K$, a stripes plot has a rectangular area that is vertically divided into K smaller rectangles, with each smaller rectangle, i, containing information about distances of the observations in cluster i from the centroid of that cluster along with the corresponding information about observations that have cluster i as their second closest cluster. The explanation of how the plot is constructed becomes more transparent if we look at an actual example. Figure 6.19 shows a stripes plot produced with that package **flexclust** (Leisch and Dimitriadou 2019) for a five-cluster solution on a set of data generated to contain five relatively distinct clusters. Looking first at the rectangle for cluster one, we see that observations in clusters two and three have the cluster one centroid as their second closest. These observations form the two other stripes within the rectangle. Observations in cluster three are further away from cluster one, but a number of observations in cluster three have a distance to the centroid of cluster one similar to those observations that belong to cluster one. Overall though, the stripes plot in Figure 6.19 suggests that the five-cluster solution matches quite well the actual structure in the data. The situation is quite different in Figure 6.20, where the stripes plot

```
R> library("flexclust")
R> library("mvtnorm")
R> set.seed(290875)
R> x <- rbind(rmvnorm(n = 20, mean = c(0, 0),
+                     sigma = diag(2)),
+             rmvnorm(n = 20, mean = c(3, 3),
+                     sigma = 0.5 * diag(2)),
+             rmvnorm(n = 20, mean = c(7, 6),
+                     sigma = 0.5 * (diag(2) + 0.25)))
R> k <- cclust(x, k = 5, save.data = TRUE)
R> plot(k, hull = FALSE, col = rep("black", 5), xlab = "x", ylab = "y")
```

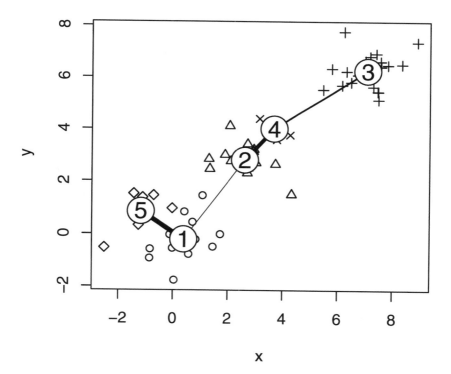

Fig. 6.17. Neighbourhood plot of *k*-means five-cluster solution for bivariate data containing three clusters.

```
R> k <- cclust(pots, k = 3, save.data = TRUE)
R> plot(k, project = prcomp(pots), hull = FALSE, col = rep("black", 3),
+       xlab = "PC1", ylab = "PC2")
```

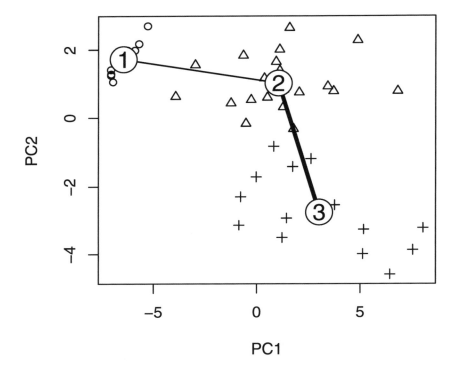

Fig. 6.18. Neighbourhood plot of k-means three-cluster solution for pottery data.

for the k-means five-group solution suggests that the clusters in this solution are not well separated, implying perhaps that the five-group solution is not appropriate for the data in this case. Lastly, the stripes plot for the k-means three-group solution on the pottery data is shown in Figure 6.21. The graphic confirms the three-group structure of the data.

All the information in a stripes plot is also available from a neighbourhood plot, but the former is dimension independent and may work well even for high-dimensional data where projections to two dimensions lose a lot of information about the structure in the data. Neither neighbourhood graphs nor stripes plots are infallible, but both offer some help in the often difficult task of evaluating and validating the solutions from a cluster analysis of a set of data.

```
R> set.seed(912345654)
R> x <- rbind(matrix(rnorm(100, sd = 0.5), ncol= 2 ),
+             matrix(rnorm(100, mean =4, sd = 0.5), ncol = 2),
+             matrix(rnorm(100, mean =7, sd = 0.5), ncol = 2),
+             matrix(rnorm(100, mean =-1.0, sd = 0.7), ncol = 2),
+             matrix(rnorm(100, mean =-4.0, sd = 1.0), ncol = 2))
R> c5 <- cclust(x, 5, save.data = TRUE)
R> stripes(c5, type = "second", col = 1)
```

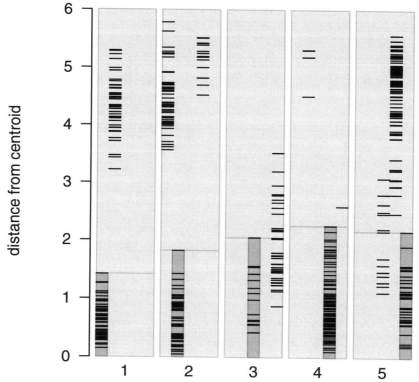

Fig. 6.19. Stripes plot of k-means solution for artificial data.

6.7 Summary

Cluster analysis techniques are used to search for clusters or groups in a priori
unclassified multivariate data. Although clustering techniques are potentially
very useful for the exploration of multivariate data, they require care in their
application if misleading solutions are to be avoided. Many methods of cluster analysis have been developed, and most studies have shown that no one
method is best for all types of data. But the more statistical techniques covered

```
R> set.seed(912345654)
R> x <- rbind(matrix(rnorm(100, sd = 2.5), ncol = 2),
+             matrix(rnorm(100, mean = 3, sd = 0.5), ncol = 2),
+             matrix(rnorm(100, mean = 5, sd = 0.5), ncol = 2),
+             matrix(rnorm(100, mean = -1.0, sd = 1.5), ncol = 2),
+             matrix(rnorm(100, mean = -4.0, sd = 2.0), ncol = 2))
R> c5 <- cclust(x, 5, save.data = TRUE)
R> stripes(c5, type = "second", col = 1)
```

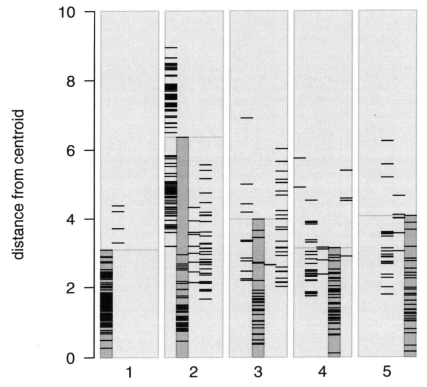

Fig. 6.20. Stripes plot of k-means solution for artificial data.

briefly in Section 6.5 and in more detail in Everitt et al. (2011) have definite *statistical* advantages because the clustering is based on sensible models for the data. Cluster analysis is a large area and has been covered only briefly in this chapter. The many problems that need to be considered when using clustering in practice have barely been touched upon. For a detailed discussion of these problems, again see Everitt et al. (2011).

```
R> set.seed(15)
R> c5 <- cclust(pots, k = 3, save.data = TRUE)
R> stripes(c5, type = "second", col = "black")
```

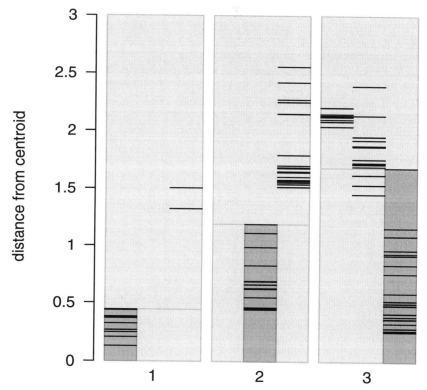

Fig. 6.21. Stripes plot of three-group k-means solution for pottery data.

Finally, we should mention in passing a technique known as projection pursuit. In essence, and like principal components analysis, projection pursuit seeks a low-dimensional projection of a multivariate data set but one that may be more likely to be successful in uncovering any cluster (or more exotic) structure in the data than principal component plots using the first few principal component scores. The technique is described in detail in Jones and Sibson (1987) and more recently in Cook and Swayne (2007).

6.8 Exercises

Ex. 6.1 Apply k-means to the crime rate data after standardising each variable by its standard deviation. Compare the results with those given in the text found by standardising by a variable's range.

Ex. 6.2 Calculate the first five principal components scores for the Romano-British pottery data, and then construct the scatterplot matrix of the scores, displaying the contours of the estimated bivariate density for each panel of the plot and a boxplot of each score in the appropriate place on the diagonal. Label the points in the scatterplot matrix with their kiln numbers.

Ex. 6.3 Return to the air pollution data given in Chapter 1 and use finite mixtures to cluster the data on the basis of the six climate and ecology variables (i.e., excluding the sulphur dioxide concentration). Investigate how sulphur dioxide concentration varies in the clusters you find both graphically and by formal significance testing.

7

Confirmatory Factor Analysis and Structural Equation Models

7.1 Introduction

An exploratory factor analysis as described in Chapter 5 is used in the early investigation of a set of multivariate data to determine whether the factor analysis model is useful in providing a parsimonious way of describing and accounting for the relationships between the observed variables. The analysis will determine which observed variables are most highly correlated with the common factors and how many common factors are needed to give an adequate description of the data. In an exploratory factor analysis, no constraints are placed on which manifest variables load on which factors. In this chapter, we will consider *confirmatory factor analysis models* in which *particular* manifest variables are allowed to relate to *particular* factors whilst other manifest variables are constrained to have zero loadings on some of the factors. A confirmatory factor analysis model may arise from theoretical considerations or be based on the results of an exploratory factor analysis where the investigator might wish to postulate a specific model for a new set of similar data, one in which the loadings of some variables on some factors are fixed at zero because they were "small" in the exploratory analysis and perhaps to allow some pairs of factors but not others to be correlated. It is important to emphasise that whilst it is perfectly appropriate to arrive at a factor model to submit to a confirmatory analysis from an exploratory factor analysis, the model *must* be tested on a fresh set of data. Models must not be generated and tested on the same data.

Confirmatory factor analysis models are a subset of a more general approach to modelling latent variables known as *structural equation modelling* or *covariance structure modelling*. Such models allow both response *and* explanatory latent variables linked by a series of linear equations. Although more complex than confirmatory factor analysis models, the aim of structural equation models is essentially the same, namely to explain the correlations or covariances of the observed variables in terms of the relationships of these variables to the assumed underlying latent variables and the relationships postulated between the latent variables themselves. Structural equation models represent the convergence of relatively independent research traditions in psychiatry, psychology, econometrics, and biometrics. The idea of latent variables in psychometrics arises from Spearman's early work on general intelligence. The concept of simultaneous directional influences of some variables on others has been part of economics for several decades, and the resulting simultaneous equation models have been used extensively by economists but essentially only with observed variables. Path analysis was introduced by Wright (1934) in a biometrics context as a method for studying the direct and indirect effects of variables. The quintessential feature of path analysis is a diagram showing how a set of explanatory variables influence a dependent variable under consideration. How the paths are drawn determines whether the explanatory variables are correlated causes, mediated causes, or independent causes. Some examples of path diagrams appear later in the chapter. (For more details of path analysis, see Schumaker and Lomax 1996).

Later, path analysis was taken up by sociologists such as Blalock (1961), Blalock (1963) and then by Duncan (1969), who demonstrated the value of combining path-analytic representation with simultaneous equation models. And, finally, in the 1970s, several workers most prominent of whom were Jöreskog (1973), Bentler (1980), and Browne (1974), combined all these various approaches into a general method that could in principle deal with extremely complex models in a routine manner.

7.2 Estimation, identification, and assessing fit for confirmatory factor and structural equation models

7.2.1 Estimation

Structural equation models will contain a number of parameters that need to be estimated from the covariance or correlation matrix of the manifest variables. Estimation involves finding values for the model parameters that minimise a *discrepancy function* indicating the magnitude of the differences between the elements of $\mathbf{S}$, the observed covariance matrix of the manifest variables and those of $\mathbf{\Sigma}(\boldsymbol{\theta})$, the covariance matrix implied by the fitted model (i.e., a matrix the elements of which are functions of the parameters of the model), contained in the vector $\boldsymbol{\theta} = (\theta_1, \ldots, \theta_t)^\top$.

There are a number of possibilities for discrepancy functions; for example, the ordinary least squares discrepancy function, FLS, is

$$\text{FLS}(\mathbf{S}, \mathbf{\Sigma}(\boldsymbol{\theta})) = \sum_{i<j} \sum_j (s_{ij} - \sigma_{ij}(\boldsymbol{\theta}))^2,$$

where s_{ij} and $\sigma_{ij}(\boldsymbol{\theta})$ are the elements of $\mathbf{S}$ and $\mathbf{\Sigma}(\boldsymbol{\theta})$. But this criterion has several problems that make it unsuitable for estimation; for example, it is not independent of the scale of the manifest variables, and so different estimates of the model parameters would be produced using the sample covariance matrix and the sample correlation matrix. Other problems with the least squares criterion are detailed in Everitt (1984).

The most commonly used method of estimating the parameters in confirmatory factor and structural equation models is maximum likelihood under the assumption that the observed data have a multivariate normal distribution. It is easy to show that maximising the likelihood is now equivalent to minimising the discrepancy function, FML, given by

$$\mathrm{FML}(\mathbf{S}, \mathbf{\Sigma}(\boldsymbol{\theta})) = \log |\mathbf{\Sigma}(\boldsymbol{\theta})| - \log |\mathbf{S}| + \mathrm{tr}(\mathbf{S}\mathbf{\Sigma}(\boldsymbol{\theta})^{-1}) - q$$

(cf. maximum likelihood factor analysis in Chapter 5). We see that by varying the parameters $\theta_1, \ldots, \theta_t$ so that $\mathbf{\Sigma}(\boldsymbol{\theta})$ becomes more like $\mathbf{S}$, FML becomes smaller. Iterative numerical algorithms are needed to minimise the function FML with respect to the parameters, but for details see Everitt (1984) and Everitt and Dunn (2001).

7.2.2 Identification

Consider the following simple example of a model in which there are three manifest variables, x, x', and y, and two latent variables, u and v, with the relationships between the manifest and latent variables being

$$x = u + \delta,$$
$$y = v + \epsilon,$$
$$x' = u + \delta'.$$

If we assume that δ, δ', and ϵ have expected values of zero, that δ and δ' are uncorrelated with each other and with u, and that ϵ is uncorrelated with v, then the covariance matrix of the three manifest variables may be expressed in terms of parameters representing the variances and covariances of the residuals and the latent variables as

$$\mathbf{\Sigma}(\boldsymbol{\theta}) = \begin{pmatrix} \theta_1 + \theta_2 & & \\ \theta_3 & \theta_4 + \theta_5 & \\ \theta_3 & \theta_4 & \theta_4 + \theta_6 \end{pmatrix},$$

where $\boldsymbol{\theta}^\top = (\theta_1, \theta_2, \theta_3, \theta_4, \theta_5, \theta_6)$ and $\theta_1 = \mathsf{Var}(v)$, $\theta_2 = \mathsf{Var}(\epsilon)$, $\theta_3 = \mathsf{Cov}(v, u)$, $\theta_4 = \mathsf{Var}(u)$, $\theta_5 = \mathsf{Var}(\delta)$, and $\theta_6 = \mathsf{Var}(\delta')$. It is immediately apparent that estimation of the parameters in this model poses a problem. The two parameters θ_1 and θ_2 are not uniquely determined because one can be, for example, increased by some amount and the other decreased by the same amount without altering the covariance matrix predicted by the model. In other words, in this example, different sets of parameter values (i.e., different $\boldsymbol{\theta}$s) will lead to the same predicted covariance matrix, $\mathbf{\Sigma}(\boldsymbol{\theta})$. The model is said to be *unidentifiable*. Formally, a model is identified if and only if $\mathbf{\Sigma}(\boldsymbol{\theta}_1) = \mathbf{\Sigma}(\boldsymbol{\theta}_2)$ implies that $\boldsymbol{\theta}_1 = \boldsymbol{\theta}_2$. In Chapter 5, it was pointed out that the parameters in the exploratory factor analysis model are not identifiable unless some constraints are introduced because different sets of factor loadings can give rise to the same predicted covariance matrix. In confirmatory factor analysis models and more general covariance structure models, identifiability depends on the choice of model and on the specification of fixed, constrained (for example, two parameters constrained to equal one another), and free parameters. If a parameter is not identified, it is not possible to find a consistent estimate of it. Establishing model identification in confirmatory factor analysis models (and in structural equation models) can be difficult because there are no simple, practicable, and universally applicable rules for evaluating whether a model is identified, although there *is* a simple necessary but not sufficient condition for identification, namely that the number of free parameters in a model, t, be less than $q(q + 1)/2$. For a more detailed discussion of the identifiability problem, see Bollen and Long (1993).

7.2.3 Assessing the fit of a model

Once a model has been pronounced identified and its parameters estimated, the next step becomes that of assessing how well the model-predicted covariance matrix fits the covariance matrix of the manifest variables. A global measure of fit of a model is provided by the likelihood ratio statistic given by $X^2 = (N - 1)\text{FML}_{\min}$, where N is the sample size and $\text{FML}_{\min}$ is the minimised value of the maximum likelihood discrepancy function given in Subsection 7.2.1. If the sample size is sufficiently large, the X^2 statistic provides a test that the population covariance matrix of the manifest variables is equal to the covariance implied by the fitted model against the alternative hypothesis that the population matrix is unconstrained. Under the equality hypothesis, X^2 has a chi-squared distribution with degrees of freedom ν given by $\frac{1}{2}q(q + 1) - t$, where t is the number of free parameters in the model.

The likelihood ratio statistic is often the only measure of fit quoted for a fitted model, but on its own it has limited practical use because in large samples even relatively trivial departures from the equality null hypothesis will lead to its rejection. Consequently, in large samples most models may be rejected as statistically untenable. A more satisfactory way to use the test is for a comparison of a series of nested models where a large difference in the statistic for two models compared with the difference in the degrees of freedom of the models indicates that the additional parameters in one of the models provide a genuine improvement in fit.

Further problems with the likelihood ratio statistic arise when the observations come from a population where the manifest variables have a non-normal distribution. Browne (1982) demonstrates that in the case of a distribution with substantial kurtosis, the chi-squared distribution may be a poor approximation for the null distribution of X^2. Browne suggests that before using the test it is advisable to assess the degree of kurtosis of the data by using

Mardia's coefficient of multivariate kurtosis (see Mardia et al. 1979). Browne's suggestion appears to be little used in practise.

Perhaps the best way to assess the fit of a model is to use the X^2 statistic alongside one or more of the following procedures:

- Visual inspection of the residual covariances (i.e., the differences between the covariances of the manifest variables and those predicted by the fitted model). These residuals should be small when compared with the values of the observed covariances or correlations.
- Examination of the standard errors of the parameters and the correlations between these estimates. If the correlations are large, it may indicate that the model being fitted is almost unidentified.
- Estimated parameter values outside their possible range; i.e., negative variances or absolute values of correlations greater than unity are often an indication that the fitted model is fundamentally wrong for the data.

In addition, a number of fit indices have been suggested that can sometimes be useful. For example, the *goodness-of-fit index* (GFI) is based on the ratio of the sum of squared distances between the matrices observed and those reproduced by the model covariance, thus allowing for scale.

The GFI measures the amount of variance and covariance in **S** that is accounted for by the covariance matrix predicted by the putative model, namely $\Sigma(\theta)$, which for simplicity we shall write as Σ. For maximum likelihood estimation, the GFI is given explicitly by

$$\text{GFI} = 1 - \frac{\text{tr}\left(\mathbf{S}\hat{\Sigma}^{-1} - \mathbf{I}\right)\left(\mathbf{S}\hat{\Sigma}^{-1} - \mathbf{I}\right)}{\text{tr}\left(\mathbf{S}\hat{\Sigma}^{-1}\mathbf{S}\hat{\Sigma}^{-1}\right)}.$$

The GFI can take values between zero (no fit) and one (perfect fit); in practise, only values above about 0.9 or even 0.95 suggest an acceptable level of fit.

The *adjusted goodness of fit index* (AGFI) adjusts the GFI index for the degrees of freedom of a model relative to the number of variables. The AGFI is calculated as follow;

$$\text{AGFI} = 1 - (k/\text{df})(1 - \text{GFI}),$$

where k is the number of unique values in **S** and df is the number of degrees of freedom in the model (discussed later). The GFI and AGFI can be used to compare the fit of two different models with the same data or compare the fit of models with different data, for example male and female data sets.

A further fit index is the *root-mean-square residual* (RMSR), which is the square root of the mean squared differences between the elements in **S** and $\hat{\Sigma}$. It can be used to compare the fit of two different models with the same data. A value of RMSR < 0.05 is generally considered to indicate a reasonable fit.

A variety of other fit indices have been proposed, including the *Tucker-Lewis index* and the *normed fit index*; for details, see Bollen and Long (1993).

7.3 Confirmatory factor analysis models

In a confirmatory factor model the loadings for some observed variables on some of the postulated common factors will be set a priori to zero. Additionally, some correlations between factors might also be fixed at zero. Such a model is fitted to a set of data by estimating its *free* parameters; i.e., those not fixed at zero by the investigator. Estimation is usually by maximum likelihood using the FML discrepancy function.

We will now illustrate the application of confirmatory factor analysis with two examples.

7.3.1 Ability and aspiration

Calsyn and Kenny (1977) recorded the values of the following six variables for 556 white eighth-grade students:

SCA: self-concept of ability;
PPE: perceived parental evaluation;
PTE: perceived teacher evaluation;
PFE: perceived friend's evaluation;
EA: educational aspiration;
CP: college plans.

Calsyn and Kenny (1977) postulated that two underlying latent variables, *ability* and *aspiration*, generated the relationships between the observed variables. The first four of the manifest variables were assumed to be indicators of ability and the last two indicators of aspiration; the latent variables, ability and aspiration, are assumed to be correlated. The regression-like equations that specify the postulated model are

$$SCA = \lambda_1 f_1 + 0 f_2 + u_1,$$
$$PPE = \lambda_2 f_1 + 0 f_2 + u_2,$$
$$PTE = \lambda_3 f_1 + 0 f_2 + u_3,$$
$$PFE = \lambda_4 f_1 + 0 f_2 + u_4,$$
$$AE = 0 f_1 + \lambda_5 f_2 + u_5,$$
$$CP = 0 f_1 + \lambda_6 f_2 + u_6,$$

where f_1 represents the ability latent variable and f_2 represents the aspiration latent variable. Note that, unlike in exploratory factor analysis, a number of factor loadings are fixed at zero and play no part in the estimation process. The model has a total of 13 parameters to estimate, six factor loadings (λ_1 to λ_6), six specific variances (ψ_1 to ψ_6), and one correlation between ability and aspiration (ρ). (To be consistent with the nomenclature used in Subsection 7.2.1, all parameters should be suffixed thetas; this could, however, become confusing, so we have changed the nomenclature and use lambdas, etc., in a manner similar to how they are used in Chapter 5.) The observed correlation matrix given in Figure 7.1 has six variances and 15 correlations, a total of 21 terms. Consequently, the postulated model has $21 - 13 = 8$ degrees of freedom. The figure depicts each correlation by an ellipse whose shape tends towards a line with slope 1 for correlations near 1, to a circle for correlations near zero, and to a line with negative slope -1 for negative correlations near -1. In addition, 100 times the correlation coefficient is printed inside the ellipse and colour-coding indicates strong negative (dark) to strong positive (light) correlations.

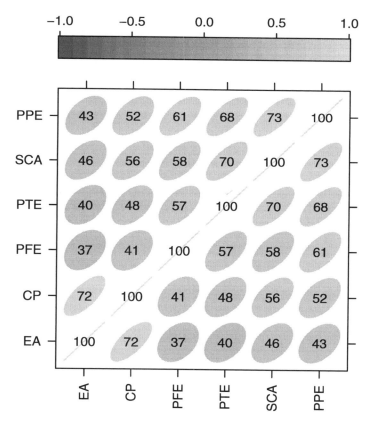

Fig. 7.1. Correlation matrix of ability and aspiration data; values given are correlation coefficients $\times 100$.

The R code, contained in the package **sem** (Fox, Kramer, and Friendly 2010), for fitting the model is

```
R> ability_model <- specify.model(file = "ability_model.txt")
R> ability_sem <- sem(ability_model, ability, 556)
```

Here, the model is specified in a text file (called `ability_model.txt` in our case) with the following content:

```
Ability     -> SCA, lambda1, NA
Ability     -> PPE, lambda2, NA
Ability     -> PTE, lambda3, NA
Ability     -> PFE, lambda4, NA
Aspiration  -> EA, lambda5, NA
Aspiration  -> CP, lambda6, NA
Ability     <-> Aspiration, rho, NA
SCA         <-> SCA, theta1, NA
PPE         <-> PPE, theta2, NA
PTE         <-> PTE, theta3, NA
PFE         <-> PFE, theta4, NA
EA          <-> EA, theta5, NA
CP          <-> CP, theta6, NA
Ability     <-> Ability, NA, 1
Aspiration  <-> Aspiration, NA, 1
```

The model is specified via arrows in the so-called reticular action model (RAM) notation. The text consists of three columns. The first one corresponds to an arrow specification where single-headed or directional arrows correspond to regression coefficients and double-headed or bidirectional arrows correspond to variance parameters. The second column denotes parameter names, and the third one assigns values to fixed parameters. Further details are available from the corresponding pages of the manual for the **sem** package.

The results from fitting the ability and aspiration model to the observed correlations are available via

R> summary(ability_sem)

```
Model Chisquare =  9.2557   Df =  8 Pr(>Chisq) = 0.32118
Chisquare (null model) =  1832.0   Df =  15
Goodness-of-fit index =  0.99443
Adjusted goodness-of-fit index =  0.98537
RMSEA index =  0.016817   90% CI: (NA, 0.05432)
Bentler-Bonnett NFI =  0.99495
Tucker-Lewis NNFI =  0.9987
Bentler CFI =  0.9993
SRMR =  0.012011
BIC =  -41.310


Normalized Residuals
   Min. 1st Qu.  Median   Mean 3rd Qu.    Max.
-0.4410 -0.1870  0.0000 -0.0131  0.2110  0.5330


Parameter Estimates
```

```
          Estimate Std Error  z value Pr(>|z|)
lambda1  0.86320  0.035182  24.5355 0.0000000
lambda2  0.84932  0.035489  23.9323 0.0000000
lambda3  0.80509  0.036409  22.1123 0.0000000
lambda4  0.69527  0.038678  17.9757 0.0000000
lambda5  0.77508  0.040365  19.2020 0.0000000
lambda6  0.92893  0.039417  23.5665 0.0000000
rho      0.66637  0.030955  21.5273 0.0000000
theta1   0.25488  0.023502  10.8450 0.0000000
theta2   0.27865  0.024263  11.4847 0.0000000
theta3   0.35184  0.026916  13.0715 0.0000000
theta4   0.51660  0.034820  14.8365 0.0000000
theta5   0.39924  0.038214  10.4475 0.0000000
theta6   0.13709  0.043530   3.1493 0.0016366


lambda1 SCA <--- Ability
lambda2 PPE <--- Ability
lambda3 PTE <--- Ability
lambda4 PFE <--- Ability
lambda5 EA <--- Aspiration
lambda6 CP <--- Aspiration
rho     Aspiration <--> Ability
theta1  SCA <--> SCA
theta2  PPE <--> PPE
theta3  PTE <--> PTE
theta4  PFE <--> PFE
theta5  EA <--> EA
theta6  CP <--> CP


Iterations =  28
```

(Note that the two latent variables have their variances fixed at one, although it is the fixing that is important, not the value at which they are fixed; these variances cannot be free parameters to be estimated.) The z values test whether parameters are significantly different from zero. All have very small associated p-values. Of particular note amongst the parameter estimates is the correlation between "true" ability and "true" aspiration; this is known as a *disattenuated correlation* and is uncontaminated by measurement error in the observed indicators of the two latent variables. In this case the estimate is 0.666, with a standard error of 0.031. An approximate 95% confidence interval for the disattenuated correlation is $[0.606, 0.727]$.

A *path diagram* (see Everitt and Dunn 2001) for the correlated, two-factor model is shown in Figure 7.2. Note that the R function `path.diagram()` "only" produces a textual representation of the graph, here in file `ability_sem.dot`.

The graphviz graph visualisation software (Gansner and North 2000) needs to be installed in order to compile the corresponding PDF file.

```
R> path.diagram(ability_sem)
```

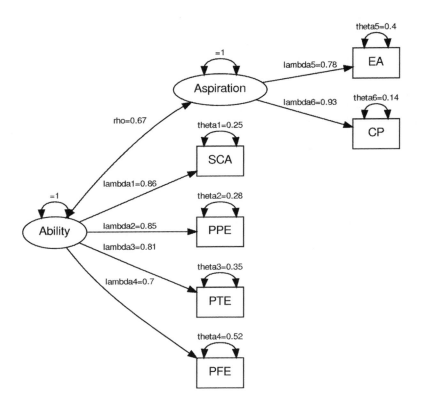

Fig. 7.2. Ability path diagram.

The fit of the model can be partially judged using the chi-square statistic described in Subsection 7.2.3, which in this case takes the value 9.256 with eight degrees of freedom and an associated p-value of 0.321, suggesting that the postulated model fits the data very well. (The chi-square test for the null model is simply a test that the population covariance matrix of the observed variables is diagonal; i.e., that the observed variables are independent. In most cases, this null model will be rejected; if it is not, the model-fitting exercise is a waste of time.) The various fit indices also indicate that the model is a good fit for the data. Also helpful in assessing the fit of the model are the summary statistics for the *normed residuals*, which are essentially the differences between corresponding elements of $\mathbf{S}$ and $\hat{\mathbf{\Sigma}}(\boldsymbol{\theta})$ but scaled so that they are unaffected by the differences in the variances of the observed variables. The normed residuals, r_{ij}^*, are defined as

$$r_{ij}^* = \frac{s_{ij} - \hat{\sigma}_{ij}}{(\hat{\sigma}_{ij}\hat{\sigma}_j^2 + \hat{\sigma}_{ij}^2)/n)^{1/2}}.$$

Generally the absolute values of the normed residuals should all be less than 2 to claim that the current model fits the data well. In the ability and aspiration example, this seems to be the case. (Note that with confirmatory factor models the standard errors of parameters become of importance because they allow the investigator to assess whether parameters might be dropped from the model to find a more parsimonious model that still provides an adequate fit to the data. In exploratory factor analysis, standard errors of factor loadings can be calculated, but they are hardly ever used; instead an informal interpretation of factors is made.)

7.3.2 A confirmatory factor analysis model for drug use

For our second example of fitting a confirmatory factor analysis model, we return to the drug use among students data introduced in Chapter 5. In the original investigation of these data reported by Huba et al. (1981), a confirmatory factor analysis model was postulated, the model arising from consideration of previously reported research on student drug use. The model involved the following three latent variables:

f_1: Alcohol use, with non-zero loadings on beer, wine, spirits, and cigarette use.

f_2: Cannabis use, with non-zero loadings on marijuana, hashish, cigarette, and wine use. The cigarette variable is assumed to load on both the first and second latent variables because it sometimes occurs with both alcohol and marijuana use and at other times does not. The non-zero loading on wine was allowed because of reports that wine is frequently used with marijuana and that consequently some of the use of wine might be an indicator of tendencies toward cannabis use.

f_3: Hard drug use, with non-zero loadings on amphetamines, tranquillizers, hallucinogenics, hashish, cocaine, heroin, drug store medication, inhalants, and spirits. The use of each of these substances was considered to suggest a strong commitment to the notion of psychoactive drug use.

Each pair of latent variables is assumed to be correlated so that these correlations are allowed to be free parameters that need to be estimated. The variance of each latent variance must, however, be fixed–they are not free parameters that can be estimated–and here as usual we will specify that each of these variances takes the value one. So the proposed model can be specified by the following series of equations:

$$\text{cigarettes} = \lambda_1 f_1 + \lambda_2 f_2 + 0 f_3 + u_1,$$
$$\text{beer} = \lambda_3 f_1 + 0 f_2 + 0 f_3 + u_2,$$
$$\text{wine} = \lambda_4 f_1 + \lambda_5 f_2 + 0 f_3 + u_3,$$
$$\text{spirits} = \lambda_6 f_1 + 0 f_2 + \lambda_7 f_3 + u_4,$$
$$\text{cocaine} = 0 f_1 + 0 f_2 + \lambda_8 f_3 + u_5,$$
$$\text{tranquillizers} = 0 f_1 + 0 f_2 + \lambda_9 f_3 + u_6,$$
$$\text{drug store medication} = 0 f_1 + 0 f_2 + \lambda_{10} f_3 + u_7,$$
$$\text{heroin} = 0 f_1 + 0 f_2 + \lambda_{11} f_3 + u_8,$$
$$\text{marijuana} = 0 f_1 + \lambda_{12} f_2 + 0 f_3 + u_9,$$
$$\text{inhalants} = 0 f_1 + 0 f_2 + \lambda_{15} f_3 + u_{11},$$
$$\text{hallucinogenics} = 0 f_1 + 0 f_2 + \lambda_{16} f_3 + u_{12},$$
$$\text{amphetamines} = 0 f_1 + 0 f_2 + \lambda_{17} f_3 + u_{13}.$$

The proposed model also allows for non-zero correlations between each pair of latent variables and so has a total of 33 parameters to estimate–17 loadings (λ_1 to λ_{17}), 13 specific variances (ψ_1 to ψ_{13}), and three correlations between latent variables (ρ_1 to ρ_3). Consequently, the model has $91 - 33 = 58$ degrees of freedom. We first abbreviate the names of the variables via

```
R> rownames(druguse) <- colnames(druguse) <- c("Cigs",
+      "Beer", "Wine", "Liqr", "Cocn", "Tran", "Drug",
+      "Hern", "Marj", "Hash", "Inhl", "Hall", "Amph")
```

To fit the model, we can use the R code

```
R> druguse_model <- specify.model(file = "druguse_model.txt")
R> druguse_sem <- sem(druguse_model, druguse, 1634)
```

where the model (stored in the text file `druguse_model.txt`) reads

```
Alcohol    -> Cigs, lambda1, NA
Alcohol    -> Beer, lambda3, NA
Alcohol    -> Wine, lambda4, NA
Alcohol    -> Liqr, lambda6, NA
Cannabis   -> Cigs, lambda2, NA
Cannabis   -> Wine, lambda5, NA
Cannabis   -> Marj, lambda12, NA
Cannabis   -> Hash, lambda13, NA
Hard       -> Liqr, lambda7, NA
Hard       -> Cocn, lambda8, NA
Hard       -> Tran, lambda9, NA
Hard       -> Drug, lambda10, NA
Hard       -> Hern, lambda11, NA
```

```
Hard        -> Hash, lambda14, NA
Hard        -> Inhl, lambda15, NA
Hard        -> Hall, lambda16, NA
Hard        -> Amph, lambda17, NA
Cigs        <-> Cigs, theta1, NA
Beer        <-> Beer, theta2, NA
Wine        <-> Wine, theta3, NA
Liqr        <-> Liqr, theta4, NA
Cocn        <-> Cocn, theta5, NA
Tran        <-> Tran, theta6, NA
Drug        <-> Drug, theta7, NA
Hern        <-> Hern, theta8, NA
Marj        <-> Marj, theta9, NA
Hash        <-> Hash, theta10, NA
Inhl        <-> Inhl, theta11, NA
Hall        <-> Hall, theta12, NA
Amph        <-> Amph, theta13, NA
Alcohol   <-> Alcohol, NA, 1
Cannabis <-> Cannabis, NA, 1
Hard        <-> Hard, NA, 1
Alcohol   <-> Cannabis, rho1, NA
Alcohol   <-> Hard, rho2, NA
Cannabis <-> Hard, rho3, NA
```

The results of fitting the proposed model are

R> summary(druguse_sem)

```
Model Chisquare =  324.09   Df =  58 Pr(>Chisq) = 0
Chisquare (null model) =  6613.7   Df =  78
Goodness-of-fit index =  0.9703
Adjusted goodness-of-fit index =  0.9534
RMSEA index =  0.053004   90% CI: (0.047455, 0.058705)
Bentler-Bonnett NFI =  0.951
Tucker-Lewis NNFI =  0.94525
Bentler CFI =  0.95929
SRMR =  0.039013
BIC =  -105.04


Normalized Residuals
  Min. 1st Qu.  Median   Mean 3rd Qu.    Max.
-3.0500 -0.8800  0.0000 -0.0217  0.9990  4.5800

Parameter Estimates
        Estimate Std Error z value Pr(>|z|)
lambda1   0.35758 0.034332  10.4153 0.0000e+00
```

```
lambda3     0.79159 0.022684   34.8962 0.0000e+00
lambda4     0.87588 0.037963   23.0716 0.0000e+00
lambda6     0.72176 0.023575   30.6150 0.0000e+00
lambda2     0.33203 0.034661    9.5793 0.0000e+00
lambda5    -0.15202 0.037155   -4.0914 4.2871e-05
lambda12    0.91237 0.030833   29.5907 0.0000e+00
lambda13    0.39549 0.030061   13.1559 0.0000e+00
lambda7     0.12347 0.022878    5.3971 6.7710e-08
lambda8     0.46467 0.025954   17.9038 0.0000e+00
lambda9     0.67554 0.024001   28.1468 0.0000e+00
lambda10    0.35842 0.026488   13.5312 0.0000e+00
lambda11    0.47591 0.025813   18.4367 0.0000e+00
lambda14    0.38199 0.029533   12.9343 0.0000e+00
lambda15    0.54297 0.025262   21.4940 0.0000e+00
lambda16    0.61825 0.024566   25.1667 0.0000e+00
lambda17    0.76336 0.023224   32.8695 0.0000e+00
theta1      0.61155 0.023495   26.0284 0.0000e+00
theta2      0.37338 0.020160   18.5210 0.0000e+00
theta3      0.37834 0.023706   15.9597 0.0000e+00
theta4      0.40799 0.019119   21.3398 0.0000e+00
theta5      0.78408 0.029381   26.6863 0.0000e+00
theta6      0.54364 0.023469   23.1644 0.0000e+00
theta7      0.87154 0.031572   27.6051 0.0000e+00
theta8      0.77351 0.029066   26.6126 0.0000e+00
theta9      0.16758 0.044839    3.7374 1.8592e-04
theta10     0.54692 0.022352   24.4691 0.0000e+00
theta11     0.70518 0.027316   25.8159 0.0000e+00
theta12     0.61777 0.025158   24.5551 0.0000e+00
theta13     0.41729 0.021422   19.4797 0.0000e+00
rho1        0.63317 0.028006   22.6079 0.0000e+00
rho2        0.31320 0.029574   10.5905 0.0000e+00
rho3        0.49893 0.027212   18.3349 0.0000e+00


lambda1  Cigs <--- Alcohol
lambda3  Beer <--- Alcohol
lambda4  Wine <--- Alcohol
lambda6  Liqr <--- Alcohol
lambda2  Cigs <--- Cannabis
lambda5  Wine <--- Cannabis
lambda12 Marj <--- Cannabis
lambda13 Hash <--- Cannabis
lambda7  Liqr <--- Hard
lambda8  Cocn <--- Hard
lambda9  Tran <--- Hard
lambda10 Drug <--- Hard
```

```
lambda11 Hern <--- Hard
lambda14 Hash <--- Hard
lambda15 Inhl <--- Hard
lambda16 Hall <--- Hard
lambda17 Amph <--- Hard
theta1    Cigs <--> Cigs
theta2    Beer <--> Beer
theta3    Wine <--> Wine
theta4    Liqr <--> Liqr
theta5    Cocn <--> Cocn
theta6    Tran <--> Tran
theta7    Drug <--> Drug
theta8    Hern <--> Hern
theta9    Marj <--> Marj
theta10   Hash <--> Hash
theta11   Inhl <--> Inhl
theta12   Hall <--> Hall
theta13   Amph <--> Amph
rho1      Cannabis <--> Alcohol
rho2      Hard <--> Alcohol
rho3      Hard <--> Cannabis

Iterations =  31
```

Here the chi-square test for goodness of fit takes the value 324.092, which with 58 degrees of freedom has an associated p-value that is very small; the model does not appear to fit very well. But before we finally decide that the fitted model is unsuitable for the data, we should perhaps investigate its fit in other ways. Here we will look at the differences of the elements of the observed covariance matrix and the covariance matrix of the fitted model. We can find these differences using the following R code:

```
R> round(druguse_sem$S - druguse_sem$C, 3)
```

```
        Cigs   Beer   Wine   Liqr   Cocn   Tran   Drug   Hern
Cigs   0.000 -0.002  0.010 -0.009 -0.015  0.015 -0.009 -0.050
Beer  -0.002  0.000  0.002  0.002 -0.047 -0.021  0.014 -0.055
Wine   0.010  0.002  0.000 -0.004 -0.039  0.005  0.039 -0.028
Liqr  -0.009  0.002 -0.004  0.000 -0.047  0.022 -0.003 -0.069
Cocn  -0.015 -0.047 -0.039 -0.047  0.000  0.035  0.042  0.100
Tran   0.015 -0.021  0.005  0.022  0.035  0.000 -0.021  0.034
Drug  -0.009  0.014  0.039 -0.003  0.042 -0.021  0.000  0.030
Hern  -0.050 -0.055 -0.028 -0.069  0.100  0.034  0.030  0.000
Marj   0.003 -0.012 -0.002  0.009 -0.026  0.007 -0.013 -0.063
Hash  -0.023  0.025  0.005  0.029  0.034 -0.014 -0.045 -0.057
Inhl   0.094  0.068  0.075  0.065  0.020 -0.044  0.115  0.030
```

```
Hall  -0.071 -0.065 -0.049 -0.077 -0.008 -0.051  0.010  0.026
Amph   0.033  0.010  0.032  0.026 -0.077  0.029 -0.042 -0.049
        Marj   Hash   Inhl   Hall   Amph
Cigs   0.003 -0.023  0.094 -0.071  0.033
Beer  -0.012  0.025  0.068 -0.065  0.010
Wine  -0.002  0.005  0.075 -0.049  0.032
Liqr   0.009  0.029  0.065 -0.077  0.026
Cocn  -0.026  0.034  0.020 -0.008 -0.077
Tran   0.007 -0.014 -0.044 -0.051  0.029
Drug  -0.013 -0.045  0.115  0.010 -0.042
Hern  -0.063 -0.057  0.030  0.026 -0.049
Marj   0.000 -0.001  0.054 -0.077  0.047
Hash  -0.001  0.000 -0.013  0.010  0.025
Inhl   0.054 -0.013  0.000  0.004 -0.022
Hall  -0.077  0.010  0.004  0.000  0.039
Amph   0.047  0.025 -0.022  0.039  0.000
```

Some of these "raw" residuals look quite large in terms of a correlational scale; for example, that corresponding to drug store medication and inhalants. And the summary statistics for the normalised residuals show that the largest is far greater than the acceptable value of 2 and the smallest is rather less than the acceptable value of -2. Perhaps the overall message for the goodness-of-fit measures is that the fitted model does not provide an entirely adequate fit for the relationships between the observed variables. Readers are referred to the original paper by Huba et al. (1981) for details of how the model was changed to try to achieve a better fit.

7.4 Structural equation models

Confirmatory factor analysis models are relatively simple examples of a more general framework for modelling latent variables and are known as either structural equation models or covariance structure models. In such models, observed variables are again assumed to be indicators of underlying latent variables, but now regression equations linking the latent variables are incorporated. Such models are fitted as described in Subsection 7.2.1. We shall illustrate these more complex models by way of a single example.

7.4.1 Stability of alienation

To illustrate the fitting of a structural equation model, we shall look at a study reported by Wheaton, Muthen, Alwin, and Summers (1977) concerned with the stability over time of attitudes such as alienation and the relationship of such attitudes to background variables such as education and occupation. For this purpose, data on attitude scales were collected from 932 people in two rural regions in Illinois at three time points, 1966, 1967, and 1971. Here we

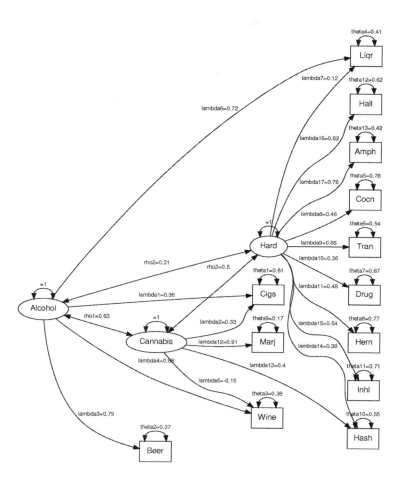

Fig. 7.3. Drug use path diagram.

shall only consider the data from 1967 and 1971. Scores on the anomia scale and powerlessness scale were taken to be indicators of the assumed latent variable, *alienation*. A respondent's years of schooling (education) and Duncan's socioeconomic index were assumed to be indicators of a respondent's *socioeconomic status*. The correlation matrix for the six observed variables is shown in Figure 7.4. The path diagram for the model to be fitted is shown in Figure 7.5. The latent variable socioeconomic status is considered to affect alienation at both time points, and alienation in 1967 also affects alienation in 1971.

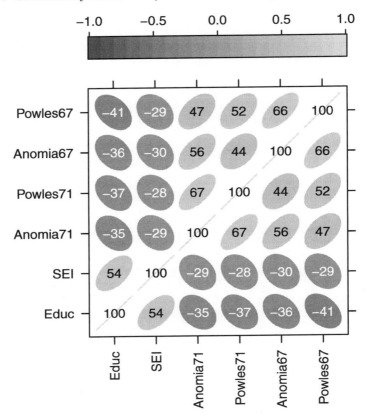

Fig. 7.4. Correlation matrix of alienation data; values given are correlation coefficients ×100.

The scale of the three latent variables, SES, alienation 67, and alienation 71, are arbitrary and have to be fixed in some way to make the model identifiable. Here each is set to the scale of one of its indicator variables by fixing the corresponding regression coefficient to one. Consequently, the equations defining the model to be fitted are as follows:

$$\text{Education} = \text{SES} + u_1,$$
$$\text{SEI} = \lambda_1 \text{SES} + u_2,$$
$$\text{Anomia67} = \text{Alienation67} + u_3,$$
$$\text{Powerlessness67} = \lambda_2 \text{Alienation67} + u_4,$$

$$\text{Anomia71} = \text{Alienation71} + u_5,$$
$$\text{Powerlessness71} = \lambda_3\text{Alienation71} + u_6,$$
$$\text{Alienation67} = \beta_1\text{SES} + u_7,$$
$$\text{Alienation71} = \beta_2\text{SES} + \beta_3\text{Alienation67} + u_8.$$

In addition to the six regression coefficients in these equations, the model also has to estimate the variances of the eight error terms, $u_1, \ldots, u_8$, and the variance of the error term for the latent variables, SES. The necessary R code for fitting the model is

```
R> alienation_model <- specify.model(
+        file = "alienation_model.txt")
R> alienation_sem <- sem(alienation_model, alienation, 932)
```

where the model reads

```
SES             -> Educ, NA, 1
SES             -> SEI, lambda1, NA
Alienation67    -> Anomia67, NA, 1
Alienation67    -> Powles67, lambda2, NA
Alienation71    -> Anomia71, NA, 1
Alienation71    -> Powles71, lambda3, NA
SES             -> Alienation67, beta1, NA
SES             -> Alienation71, beta2, NA
Alienation67    -> Alienation71, beta3, NA
Educ          <-> Educ, theta1, NA
SEI           <-> SEI, theta2, NA
SES           <-> SES, delta0, NA
Anomia67      <-> Anomia67, theta3, NA
Powles67      <-> Powles67, theta4, NA
Anomia71      <-> Anomia71, theta5, NA
Powles71      <-> Powles71, theta6, NA
Alienation67  <-> Alienation67, delta1, NA
Alienation71  <-> Alienation71, delta2, NA
```

The parameter estimates are

```
R> summary(alienation_sem)

 Model Chisquare =   71.532   Df =   6 Pr(>Chisq) = 1.9829e-13
 Chisquare (null model) =  2131.5   Df =   15
 Goodness-of-fit index =   0.97514
 Adjusted goodness-of-fit index =   0.913
 RMSEA index =  0.10831   90% CI: (0.086636, 0.13150)
 Bentler-Bonnett NFI =   0.96644
 Tucker-Lewis NNFI =   0.9226
 Bentler CFI =   0.96904
```

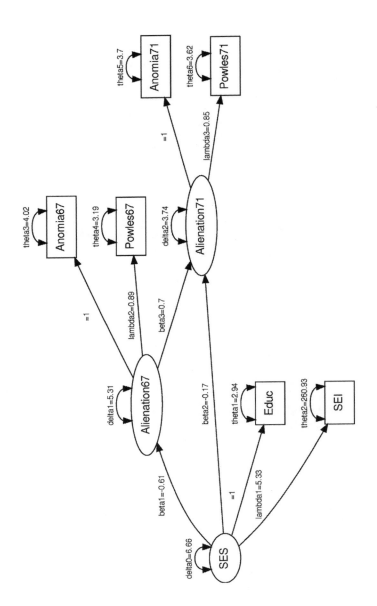

Fig. 7.5. Alienation path diagram.

```
SRMR =   0.021256
BIC =   30.508
```

Normalized Residuals
```
  Min. 1st Qu.  Median     Mean 3rd Qu.    Max.
-1.2600 -0.2090 -0.0001 -0.0151  0.2440  1.3300
```

Parameter Estimates

	Estimate	Std Error	z value	Pr(>|z|)
lambda1	5.33054	0.430948	12.3693	0.0000e+00
lambda2	0.88883	0.043229	20.5609	0.0000e+00
lambda3	0.84892	0.041567	20.4229	0.0000e+00
beta1	-0.61361	0.056262	-10.9063	0.0000e+00
beta2	-0.17447	0.054221	-3.2178	1.2920e-03
beta3	0.70463	0.053387	13.1984	0.0000e+00
theta1	2.93614	0.500979	5.8608	4.6064e-09
theta2	260.93220	18.275902	14.2774	0.0000e+00
delta0	6.66392	0.641907	10.3814	0.0000e+00
theta3	4.02305	0.359231	11.1990	0.0000e+00
theta4	3.18933	0.284033	11.2288	0.0000e+00
theta5	3.70315	0.391837	9.4508	0.0000e+00
theta6	3.62334	0.304359	11.9048	0.0000e+00
delta1	5.30685	0.484186	10.9603	0.0000e+00
delta2	3.73998	0.388683	9.6222	0.0000e+00

```
lambda1 SEI <--- SES
lambda2 Powles67 <--- Alienation67
lambda3 Powles71 <--- Alienation71
beta1   Alienation67 <--- SES
beta2   Alienation71 <--- SES
beta3   Alienation71 <--- Alienation67
theta1  Educ <--> Educ
theta2  SEI <--> SEI
delta0  SES <--> SES
theta3  Anomia67 <--> Anomia67
theta4  Powles67 <--> Powles67
theta5  Anomia71 <--> Anomia71
theta6  Powles71 <--> Powles71
delta1  Alienation67 <--> Alienation67
delta2  Alienation71 <--> Alienation71
```

```
Iterations =  84
```

The value of the chi-square fit statistic is 71.532, which with 6 degrees of freedom suggests that the model does not fit well. Jöreskog and Sörbom (1981) suggest that the model can be improved by allowing the measurement errors for anomia in 1967 and in 1971 to be correlated. Fitting such a model in R requires the addition of the following line to the code above:

```
Anomia67 <-> Anomia71, psi, NA
```

The chi-square fit statistic is now 6.359 with 5 degrees of freedom. Clearly the introduction of correlated measurement errors for the two measurements of anomia has greatly improved the fit of the model. However, Bentler (1982), in a discussion of this example, suggests that the importance of the structure remaining to be explained after fitting the original model is in practical terms very small, and Browne (1982) criticises the tendency to allow error terms to become correlated simply to obtain an improvement in fit unless there are sound theoretical reasons why particular error terms should be related.

7.5 Summary

The possibility of making causal inferences about latent variables is one that has great appeal, particularly for social and behavioural scientists, simply because the concepts in which they are most interested are rarely measurable directly. And because such models can nowadays be relatively easily fitted, researchers can routinely investigate quite complex models. But perhaps a caveat issued more than 20 years ago still has some relevance–the following is from Cliff (1983):

> Correlational data are still correlational and no computer program can take account of variables that are not in the analysis. Causal relations can only be established through patient, painstaking attention to all the relevant variables, and should involve active manipulation as a final confirmation.

The maximum likelihood estimation approach described in this chapter is based on the assumption of multivariate normality for the data. When a multivariate normality assumption is clearly untenable, for example with categorical variables, applying the maximum likelihood methods can lead to biased results, although when there are five or more response categories (and the distribution of the data is normal) the problems from disregarding the categorical nature of variables are likely to be minimised (see Marcoulides 2005). Muthen (1984) describes a very general approach for structural equation modelling that can be used when the data consist of a mixture of continuous, ordinal, and dichotomous variables.

Sample size issues for structural equation modelling have been considered by MacCullum, Browne, and Sugawara (1996), and Muthen and Muthen (2002) illustrate how to undertake a Monte Carlo study to help decide on sample size and determine power.

7.6 Exercises

Ex. 7.1 The matrix below shows the correlations between ratings on nine
statements about pain made by 123 people suffering from extreme pain.
Each statement was scored on a scale from 1 to 6, ranging from agreement
to disagreement. The nine pain statements were as follows:
1. Whether or not I am in pain in the future depends on the skills of the
doctors.
2. Whenever I am in pain, it is usually because of something I have done
or not done.
3. Whether or not I am in pain depends on what the doctors do for me.
4. I cannot get any help for my pain unless I go to seek medical advice.
5. When I am in pain, I know that it is because I have not been taking
proper exercise or eating the right food.
6. People's pain results from their own carelessness.
7. I am directly responsible for my pain,
8. relief from pain is chiefly controlled by the doctors.
9. People who are never in pain are just plain lucky.

$$
\begin{pmatrix}
1.00 \\
-0.04 & 1.00 \\
0.61 & -0.07 & 1.00 \\
0.45 & -0.12 & 0.59 & 1.00 \\
0.03 & 0.49 & 0.03 & -0.08 & 1.00 \\
-0.29 & 0.43 & -0.13 & -0.21 & 0.47 & 1.00 \\
-0.30 & 0.30 & -0.24 & -0.19 & 0.41 & 0.63 & 1.00 \\
0.45 & -0.31 & 0.59 & 0.63 & -0.14 & -0.13 & -0.26 & 1.00 \\
0.30 & -0.17 & 0.32 & 0.37 & -0.24 & -0.15 & -0.29 & 0.40 & 1.00
\end{pmatrix}.
$$

Fit a correlated two-factor model in which questions $1, 3, 4,$ and 8 are
assumed to be indicators of the latent variable *Doctor's Responsibility*
and questions $2, 5, 6,$ and 7 are assumed to be indicators of the latent
variable *Patient's Responsibility*. Find a 95% confidence interval for the
correlation between the two latent variables.

Ex. 7.2 For the stability of alienation example, fit the model in which the
measurement errors for anomia in 1967 and anomia in 1971 are allowed
to be correlated.

Ex. 7.3 Meyer and Bendig (1961) administered the five Thurstone Primary
Mental Ability tests, verbal meaning (V), space (S), reasoning (R), numerical (N), and word fluency (W), to 49 boys and 61 girls in grade 8 and
again three and a half years later in grade 11. The observed correlation
matrix is shown below. Fit a single-factor model to the correlations that
allows the factor at time one to be correlated with the factor at time two.

$$
\begin{pmatrix}
 & V & S & R & N & W & V & S & R & N & W \\
V & 1.00 \\
S & 0.37 & 1.00 \\
R & 0.42 & 0.33 & 1.00 \\
N & 0.53 & 0.14 & 0.38 & 1.00 \\
W & 0.38 & 0.10 & 0.20 & 0.24 & 1.00 \\
V & 0.81 & 0.34 & 0.49 & 0.58 & 0.32 & 1.00 \\
S & 0.35 & 0.65 & 0.20 & -0.04 & 0.11 & 0.34 & 1.00 \\
R & 0.42 & 0.32 & 0.75 & 0.46 & 0.26 & 0.46 & 0.18 & 1.00 \\
N & 0.40 & 0.14 & 0.39 & 0.73 & 0.19 & 0.55 & 0.06 & 0.54 & 1.00 \\
W & 0.24 & 0.15 & 0.17 & 0.15 & 0.43 & 0.24 & 0.15 & 0.20 & 0.16 & 1.00
\end{pmatrix}
$$

8

The Analysis of Repeated Measures Data

8.1 Introduction

The multivariate data sets considered in previous chapters have involved measurements or observations on a number of different variables for each object or individual in the study. In this chapter, however, we will consider multivariate data of a different nature, namely data resulting from the repeated measurements of the same variable on each unit in the data set. Examples of such data are common in many disciplines. But before we introduce some actual repeated measures data sets, we need to make a small digression in order to introduce the two different "formats", the wide and the long forms, in which such data are commonly stored and dealt with. The simplest way to do this is by way of a small example data set:

```
R> ex_wide
```

	ID	Group	Day.1	Day.2	Day.5	Day.7
1	1	1	15	15	10	7
2	2	1	10	9	11	12
3	3	1	8	7	6	9
4	4	2	11	8	13	7
5	5	2	11	12	11	11
6	6	2	12	12	6	10

We can pretend that these data come from a clinical trial in which individuals have been assigned to two treatment groups and have a response variable of interest recorded on days 1, 2, 5 and 7. As given in this table, the data are in their wide form (as are most of the data sets met in earlier chapters); each row of data corresponds to an individual and contains all the repeated measures for the individual as well as other variables that might have been recorded–in this case the treatment group of the individual. These data can be put into their long form, in which each row of data now corresponds to one of the repeated measurements along with the values of other variables associated with this particular measurement (for example, the time the measurement was taken) by using the reshape() function in R. The code needed to rearrange the data ex_wide into their long form is as follows:

```R
R> reshape(ex_wide, direction = "long", idvar = "ID",
+          varying = colnames(ex_wide)[-(1:2)])
```

	ID	Group	time	Day
1.1	1	1	1	15
2.1	2	1	1	10
3.1	3	1	1	8
4.1	4	2	1	11
5.1	5	2	1	11
6.1	6	2	1	12
1.2	1	1	2	15
2.2	2	1	2	9
3.2	3	1	2	7
4.2	4	2	2	8
5.2	5	2	2	12
6.2	6	2	2	12
1.5	1	1	5	10
2.5	2	1	5	11
3.5	3	1	5	6
4.5	4	2	5	13
5.5	5	2	5	11
6.5	6	2	5	6
1.7	1	1	7	7
2.7	2	1	7	12
3.7	3	1	7	9
4.7	4	2	7	7
5.7	5	2	7	11
6.7	6	2	7	10

Here, varying contains the names of the variables containing the repeated measurements. Note that the name of the variable consists of the name itself and the time point, separated by a dot. This long form of repeated measures data is used when applying the models to be described in Section 8.2, although the wide format is often more convenient for plotting the data and computing summary statistics.

So now let us take a look at two repeated measurement data sets that we shall be concerned with in this chapter. The first, shown in its long form in Table 8.1, is taken from Crowder (1998) and gives the loads required to produce slippage x of a timber specimen in a clamp. There are eight specimens each with 15 repeated measurements. The second data set, in Table 8.2 reported in Zerbe (1979) and also given in Davis (2003), consists of plasma inorganic phosphate measurements obtained from 13 control and 20 obese patients $0, 0.5, 1, 1.5, 2,$ and 3 hours after an oral glucose challenge. These two data sets illustrate that although repeated measurements often arise from the passing of time (longitudinal data), this is not always the case.

Table 8.1: `timber` data. Data giving loads needed for a given slippage in eight specimens of timber, with data in "long" form.

specimen	slippage	loads	specimen	slippage	loads
spec1	0.0	0.00	spec5	0.7	12.25
spec2	0.0	0.00	spec6	0.7	12.85
spec3	0.0	0.00	spec7	0.7	12.13
spec4	0.0	0.00	spec8	0.7	12.21
spec5	0.0	0.00	spec1	0.8	12.98
spec6	0.0	0.00	spec2	0.8	14.08
spec7	0.0	0.00	spec3	0.8	13.18
spec8	0.0	0.00	spec4	0.8	12.23
spec1	0.1	2.38	spec5	0.8	13.35
spec2	0.1	2.69	spec6	0.8	13.83
spec3	0.1	2.85	spec7	0.8	13.15
spec4	0.1	2.46	spec8	0.8	13.16
spec5	0.1	2.97	spec1	0.9	13.94
spec6	0.1	3.96	spec2	0.9	14.66
spec7	0.1	3.17	spec3	0.9	14.12
spec8	0.1	3.36	spec4	0.9	13.24
spec1	0.2	4.34	spec5	0.9	14.54
spec2	0.2	4.75	spec6	0.9	14.85
spec3	0.2	4.89	spec7	0.9	14.09
spec4	0.2	4.28	spec8	0.9	14.05
spec5	0.2	4.68	spec1	1.0	14.74
spec6	0.2	6.46	spec2	1.0	15.37
spec7	0.2	5.33	spec3	1.0	15.09
spec8	0.2	5.45	spec4	1.0	14.19
spec1	0.3	6.64	spec5	1.0	15.53
spec2	0.3	7.04	spec6	1.0	15.79
spec3	0.3	6.61	spec7	1.0	15.11
spec4	0.3	5.88	spec8	1.0	14.96
spec5	0.3	6.66	spec1	1.2	16.13
spec6	0.3	8.14	spec2	1.2	16.89
spec7	0.3	7.14	spec3	1.2	16.68
spec8	0.3	7.08	spec4	1.2	16.07
spec1	0.4	8.05	spec5	1.2	17.38
spec2	0.4	9.20	spec6	1.2	17.39
spec3	0.4	8.09	spec7	1.2	16.69
spec4	0.4	7.43	spec8	1.2	16.24
spec5	0.4	8.11	spec1	1.4	17.98
spec6	0.4	9.35	spec2	1.4	17.78

Table 8.1: `timber` data (continued).

specimen	slippage	loads	specimen	slippage	loads
spec7	0.4	8.29	spec3	1.4	17.94
spec8	0.4	8.32	spec4	1.4	17.43
spec1	0.5	9.78	spec5	1.4	18.76
spec2	0.5	10.94	spec6	1.4	18.44
spec3	0.5	9.72	spec7	1.4	17.69
spec4	0.5	8.32	spec8	1.4	17.34
spec5	0.5	9.64	spec1	1.6	19.52
spec6	0.5	10.72	spec2	1.6	18.41
spec7	0.5	9.86	spec3	1.6	18.22
spec8	0.5	9.91	spec4	1.6	18.36
spec1	0.6	10.97	spec5	1.6	19.81
spec2	0.6	12.23	spec6	1.6	19.46
spec3	0.6	11.03	spec7	1.6	18.71
spec4	0.6	9.92	spec8	1.6	18.23
spec5	0.6	11.06	spec1	1.8	19.97
spec6	0.6	11.84	spec2	1.8	18.97
spec7	0.6	11.07	spec3	1.8	19.40
spec8	0.6	11.06	spec4	1.8	18.93
spec1	0.7	12.05	spec5	1.8	20.62
spec2	0.7	13.19	spec6	1.8	20.05
spec3	0.7	12.14	spec7	1.8	19.54
spec4	0.7	11.10	spec8	1.8	18.87

Table 8.2: `plasma` data. Plasma inorganic phosphate levels from 33 subjects, with data in "long" form.

Subject	group	time	plasma	Subject	group	time	plasma
id01	control	1	4.3	id01	control	5	2.2
id02	control	1	3.7	id02	control	5	2.9
id03	control	1	4.0	id03	control	5	2.9
id04	control	1	3.6	id04	control	5	2.9
id05	control	1	4.1	id05	control	5	3.6
id06	control	1	3.8	id06	control	5	3.8
id07	control	1	3.8	id07	control	5	3.1
id08	control	1	4.4	id08	control	5	3.6
id09	control	1	5.0	id09	control	5	3.3
id10	control	1	3.7	id10	control	5	1.5
id11	control	1	3.7	id11	control	5	2.9
id12	control	1	4.4	id12	control	5	3.7

Table 8.2: `plasma` data (continued).

Subject	group	time	plasma	Subject	group	time	plasma
id13	control	1	4.7	id13	control	5	3.2
id14	control	1	4.3	id14	control	5	2.2
id15	control	1	5.0	id15	control	5	3.7
id16	control	1	4.6	id16	control	5	3.7
id17	control	1	4.3	id17	control	5	3.1
id18	control	1	3.1	id18	control	5	2.6
id19	control	1	4.8	id19	control	5	2.2
id20	control	1	3.7	id20	control	5	2.9
id21	obese	1	5.4	id21	obese	5	2.8
id22	obese	1	3.0	id22	obese	5	2.1
id23	obese	1	4.9	id23	obese	5	3.7
id24	obese	1	4.8	id24	obese	5	4.7
id25	obese	1	4.4	id25	obese	5	3.5
id26	obese	1	4.9	id26	obese	5	3.3
id27	obese	1	5.1	id27	obese	5	3.4
id28	obese	1	4.8	id28	obese	5	4.1
id29	obese	1	4.2	id29	obese	5	3.3
id30	obese	1	6.6	id30	obese	5	4.3
id31	obese	1	3.6	id31	obese	5	2.1
id32	obese	1	4.5	id32	obese	5	2.4
id33	obese	1	4.6	id33	obese	5	3.8
id01	control	2	3.3	id01	control	6	2.5
id02	control	2	2.6	id02	control	6	3.2
id03	control	2	4.1	id03	control	6	3.1
id04	control	2	3.0	id04	control	6	3.9
id05	control	2	3.8	id05	control	6	3.4
id06	control	2	2.2	id06	control	6	3.6
id07	control	2	3.0	id07	control	6	3.4
id08	control	2	3.9	id08	control	6	3.8
id09	control	2	4.0	id09	control	6	3.6
id10	control	2	3.1	id10	control	6	2.3
id11	control	2	2.6	id11	control	6	2.2
id12	control	2	3.7	id12	control	6	4.3
id13	control	2	3.1	id13	control	6	4.2
id14	control	2	3.3	id14	control	6	2.5
id15	control	2	4.9	id15	control	6	4.1
id16	control	2	4.4	id16	control	6	4.2
id17	control	2	3.9	id17	control	6	3.1
id18	control	2	3.1	id18	control	6	1.9
id19	control	2	5.0	id19	control	6	3.1
id20	control	2	3.1	id20	control	6	3.6
id21	obese	2	4.7	id21	obese	6	3.7

Table 8.2: `plasma` data (continued).

Subject	group	time	plasma	Subject	group	time	plasma
id22	obese	2	2.5	id22	obese	6	2.6
id23	obese	2	5.0	id23	obese	6	4.1
id24	obese	2	4.3	id24	obese	6	3.7
id25	obese	2	4.2	id25	obese	6	3.4
id26	obese	2	4.3	id26	obese	6	4.1
id27	obese	2	4.1	id27	obese	6	4.2
id28	obese	2	4.6	id28	obese	6	4.0
id29	obese	2	3.5	id29	obese	6	3.1
id30	obese	2	6.1	id30	obese	6	3.8
id31	obese	2	3.4	id31	obese	6	2.4
id32	obese	2	4.0	id32	obese	6	2.3
id33	obese	2	4.4	id33	obese	6	3.6
id01	control	3	3.0	id01	control	7	3.4
id02	control	3	2.6	id02	control	7	3.1
id03	control	3	3.1	id03	control	7	3.9
id04	control	3	2.2	id04	control	7	3.8
id05	control	3	2.1	id05	control	7	3.6
id06	control	3	2.0	id06	control	7	3.0
id07	control	3	2.4	id07	control	7	3.5
id08	control	3	2.8	id08	control	7	4.0
id09	control	3	3.4	id09	control	7	4.0
id10	control	3	2.9	id10	control	7	2.7
id11	control	3	2.6	id11	control	7	3.1
id12	control	3	3.1	id12	control	7	3.9
id13	control	3	3.2	id13	control	7	3.7
id14	control	3	3.0	id14	control	7	2.4
id15	control	3	4.1	id15	control	7	4.7
id16	control	3	3.9	id16	control	7	4.8
id17	control	3	3.1	id17	control	7	3.6
id18	control	3	3.3	id18	control	7	2.3
id19	control	3	2.9	id19	control	7	3.5
id20	control	3	3.3	id20	control	7	4.3
id21	obese	3	3.9	id21	obese	7	3.5
id22	obese	3	2.3	id22	obese	7	3.2
id23	obese	3	4.1	id23	obese	7	4.7
id24	obese	3	4.7	id24	obese	7	3.6
id25	obese	3	4.2	id25	obese	7	3.8
id26	obese	3	4.0	id26	obese	7	4.2
id27	obese	3	4.6	id27	obese	7	4.4
id28	obese	3	4.6	id28	obese	7	3.8
id29	obese	3	3.8	id29	obese	7	3.5
id30	obese	3	5.2	id30	obese	7	4.2

Table 8.2: `plasma` data (continued).

Subject	group	time	plasma	Subject	group	time	plasma
id31	obese	3	3.1	id31	obese	7	2.5
id32	obese	3	3.7	id32	obese	7	3.1
id33	obese	3	3.8	id33	obese	7	3.8
id01	control	4	2.6	id01	control	8	4.4
id02	control	4	1.9	id02	control	8	3.9
id03	control	4	2.3	id03	control	8	4.0
id04	control	4	2.8	id04	control	8	4.0
id05	control	4	3.0	id05	control	8	3.7
id06	control	4	2.6	id06	control	8	3.5
id07	control	4	2.5	id07	control	8	3.7
id08	control	4	2.1	id08	control	8	3.9
id09	control	4	3.4	id09	control	8	4.3
id10	control	4	2.2	id10	control	8	2.8
id11	control	4	2.3	id11	control	8	3.9
id12	control	4	3.2	id12	control	8	4.8
id13	control	4	3.3	id13	control	8	4.3
id14	control	4	2.6	id14	control	8	3.4
id15	control	4	3.7	id15	control	8	4.9
id16	control	4	3.9	id16	control	8	5.0
id17	control	4	3.1	id17	control	8	4.0
id18	control	4	2.6	id18	control	8	2.7
id19	control	4	2.8	id19	control	8	3.6
id20	control	4	2.8	id20	control	8	4.4
id21	obese	4	4.1	id21	obese	8	3.7
id22	obese	4	2.2	id22	obese	8	3.5
id23	obese	4	3.7	id23	obese	8	4.9
id24	obese	4	4.6	id24	obese	8	3.9
id25	obese	4	3.4	id25	obese	8	4.0
id26	obese	4	4.0	id26	obese	8	4.3
id27	obese	4	4.1	id27	obese	8	4.9
id28	obese	4	4.4	id28	obese	8	3.8
id29	obese	4	3.6	id29	obese	8	3.9
id30	obese	4	4.1	id30	obese	8	4.8
id31	obese	4	2.8	id31	obese	8	3.5
id32	obese	4	3.3	id32	obese	8	3.3
id33	obese	4	3.8	id33	obese	8	3.8

The distinguishing feature of a repeated measures study is that the response variable of interest has been recorded several times on each unit in the data set. In addition, a set of explanatory variables (covariates is an alternative term that is often used in this context) are available for each; some of the explanatory variables may have been recorded only once for each unit and so take the same value for each of the repeated response values for that unit–an example would be treatment group in a clinical trial. Other explanatory variables may take different values for each of the different response variable values–an example would be age; these are sometimes called *time-varying co-variates*.

The main objective in such a study is to characterise change in the repeated values of the response variable and to determine the explanatory variables most associated with any change. Because several observations of the response variable are made on the same individual, it is likely that the measurements will be correlated rather than independent, even after conditioning on the explanatory variables. Consequently, repeated measures data require special methods of analysis, and models for such data need to include parameters linking the explanatory variables to the repeated measurements, parameters analogous to those in the usual multiple regression model, and, in addition, parameters that account for the correlational structure of the repeated measurements. It is the former parameters that are generally of most interest, with the latter often being regarded as *nuisance parameters*. But providing an adequate model for the correlational structure of the repeated measures is necessary to avoid misleading inferences about the parameters that are of most importance to the researcher.

Over the last decade, methodology for the analysis of repeated measures data has been the subject of much research and development, and there are now a variety of powerful techniques available. Comprehensive accounts of these methods are given in Diggle, Heagerty, Liang, and Zeger (2003), Davis (2003), and Skrondal and Rabe-Hesketh (2004). Here we will concentrate on a single class of methods, *linear mixed-effects models*.

8.2 Linear mixed-effects models for repeated measures data

Linear mixed-effects models for repeated measures data formalise the sensible idea that an individual's pattern of responses is likely to depend on many characteristics of that individual, including some that are unobserved. These unobserved variables are then included in the model as random variables, that is, *random effects*. The essential feature of the model is that correlation amongst the repeated measurements on the same unit arises from shared, unobserved variables. Conditional on the values of the random effects, the repeated measurements are assumed to be independent, the so-called *local independence* assumption.

Linear mixed-effects models are introduced in the next subsection by describing two commonly used models, the *random intercept* and *random intercept and slope* models, in the context of the timber slippage data in Table 8.1.

8.2.1 Random intercept and random intercept and slope models for the timber slippage data

Let y_{ij} represent the load in specimen i needed to produce a slippage of x_j, with $i = 1, \ldots, 8$ and $j = 1, \ldots, 15$. If we choose to ignore the repeated measures structure of the, data we could fit a simple linear regression model of the form

$$y_{ij} = \beta_0 + \beta_1 x_j + \epsilon_{ij}. \tag{8.1}$$

The model in Equation (8.1) can be fitted using the long form of the data in association with the `lm()` function in R as follows:

```
R> summary(lm(loads ~ slippage, data = timber))

Call:
lm(formula = loads ~ slippage, data = timber)

Residuals:
   Min    1Q Median    3Q   Max
-3.516 -0.981  0.408  1.298  2.491

Coefficients:
            Estimate Std. Error t value Pr(>|t|)
(Intercept)    3.516      0.264    13.3   <2e-16
slippage      10.373      0.283    36.6   <2e-16

Residual standard error: 1.65 on 118 degrees of freedom
Multiple R-squared: 0.919,         Adjusted R-squared: 0.918
F-statistic: 1.34e+03 on 1 and 118 DF,  p-value: <2e-16
```

We see that the slippage effect is large and highly significant.

But such a model assumes that the repeated observations are independent of one another which is totally unrealistic for most repeated measures data sets. A more realistic model is the *random intercept model*, where by partitioning the total residual that is present in the usual linear regression model (8.1) into a *subject-specific random component* u_i that is constant over slippage plus an error term ϵ_{ij} that varies randomly over slippage, some correlational structure for the repeated measures is introduced. The random intercept model is

$$y_{ij} = (\beta_0 + u_i) + \beta_1 x_j + \epsilon_{ij}. \tag{8.2}$$

(The error terms and parameters in Equations (8.1) and (8.2) are, of course, not the same.) The u_i are assumed to be normally distributed with zero mean and variance σ_u^2, and the ϵ_{ij} are assumed normally distributed with zero mean and variance σ^2. The u_i and the ϵ_{ij} are assumed to be independent of each other and the x_j. The subject-specific random effects allow for differences in the intercepts of each individual's regression fit of load on slippage. The repeated measurements for each timber specimen vary about that specimen's *own* regression line, and this can differ in intercept but *not* in slope from the regression lines of other specimens. In the random-effects model, there is possible heterogeneity in the intercepts of the individuals. In this model, slippage has a *fixed effect*.

How does the random intercept model introduce a correlational structure for the repeated measurements? First, the random intercept model implies that the total variance of each repeated measurement is

$$\mathsf{Var}(u_i + \epsilon_{ij}) = \sigma_u^2 + \sigma^2.$$

(Due to this decomposition of the total residual variance into a between-subject component, σ_u^2, and a within-subject component, σ^2, the model is sometimes referred to as a *variance component model*.) The covariance between the total residuals at two slippage levels j and j' in the same specimen i is

$$\mathsf{Cov}(u_i + \epsilon_{ij}, u_i + \epsilon_{ij'}) = \sigma_u^2.$$

The covariance will be non-zero if the variance of the subject-specific random effects is non-zero. (Note that this covariance is induced by the shared random intercept; for specimens with $u_i > 0$, the total residuals will tend to be greater than the mean, and for specimens with $u_i < 0$ they will tend to be less than the mean.) It follows from the two relations above that the residual correlation (i.e., the correlation between pairs of repeated measurements) is given by

$$\mathsf{Cor}(u_i + \epsilon_{ij}, u_i + \epsilon_{ij'}) = \frac{\sigma_u^2}{\sigma_u^2 + \sigma^2}.$$

This is an *intra-class correlation* that is interpreted as the proportion of the total residual variance that is due to residual variability between subjects. So a random intercept model constrains the variance of each repeated measure to be the same and the covariance between any pair of repeated measurements to be equal. This is usually called the *compound symmetry* structure. These constraints often are not realistic for repeated measures data. For example, for longitudinal data, it is more common for measures taken closer to each other in time to be more highly correlated than those taken further apart. In addition, the variances of the later repeated measures are often greater than those taken earlier. Consequently, for many such data sets, the random intercept model will not do justice to the observed pattern of covariances between the repeated measures.

A model that allows a more realistic structure for the covariances is the *random slope and intercept model* that provides for heterogeneity in both slopes and intercepts.

In this model, there are two types of random effects, the first modelling heterogeneity in intercepts, u_{i1}, and the second modelling heterogeneity in slopes, u_{i2}. Explicitly, the model is

$$y_{ij} = (\beta_0 + u_{i1}) + (\beta_1 + u_{i2})x_j + \epsilon_{ij}, \tag{8.3}$$

where the parameters are not, of course, the same as in (8.1). The two random effects are assumed to have a bivariate normal distribution with zero means for both variables, variances $\sigma_{u_1}^2$ and $\sigma_{u_2}^2$, and covariance $\sigma_{u_1 u_2}$. With this model, the total residual is $u_{i1} + u_{i2}x_j + \epsilon_{ij}$ with variance

$$\mathsf{Var}(u_{i1} + u_{i2}x_j + \epsilon_{ij}) = \sigma_{u_1}^2 + 2\sigma_{u_1 u_2}x_j + \sigma_{u_2}^2 x_j^2 + \sigma^2,$$

which is no longer constant for different values of x_j. Similarly, the covariance between two total residuals of the same individual,

$$\mathsf{Cov}(u_{i1} + x_j u_{i2} + \epsilon_{ij}, u_{i1} + x_{j'}u_{i2} + \epsilon_{ij'}) = \sigma_{u_1}^2 + \sigma_{u_1 u_2}(x_j + x_{j'}) + \sigma_{u_2}^2 x_j x_{j'},$$

is not constrained to be the same for all pairs j and j'. The random intercept and slope model allows for the repeated measurements to have different variances and for pairs of repeated measurements to have different correlations.

Linear mixed-effects models can be estimated by maximum likelihood. However, this method tends to underestimate the variance components. A modified version of maximum likelihood, known as *restricted maximum likelihood*, is therefore often recommended; this provides consistent estimates of the variance components. Details are given in Diggle et al. (2003), Longford (1993), and Skrondal and Rabe-Hesketh (2004).

Competing linear mixed-effects models can be compared using a likelihood ratio test. If, however, the models have been estimated by restricted maximum likelihood, this test can only be used if both models have the same set of fixed effects (Longford 1993).

8.2.2 Applying the random intercept and the random intercept and slope models to the timber slippage data

Before beginning any formal model-fitting exercise, it is good data analysis practise to look at some informative graphic (or graphics) of the data. Here we first produce a plot of the trajectories of each timber specimen over the slippage levels; see Figure 8.1.

The figure shows that loads required to achieve a given slippage level increase with slippage value, with the increase gradually leveling off; this explains the large slippage effect found earlier when applying simple linear regression to the data.

We can now fit the two linear mixed-effects models (8.2) and (8.3) as described in the previous subsection using the lme() function from the package **nlme** (Pinheiro, Bates, DebRoy, Sarkar, and R Development Core Team 2010):

```
R> timber.lme <- lme(loads ~ slippage,
+                    random = ~1 | specimen,
+                    data = timber, method = "ML")
R> timber.lme1 <- lme(loads ~ slippage,
+                     random = ~slippage | specimen,
+                     data = timber, method = "ML")
```

```
R> xyplot(loads ~ slippage | specimen, data = timber,
+         layout = c(4, 2))
```

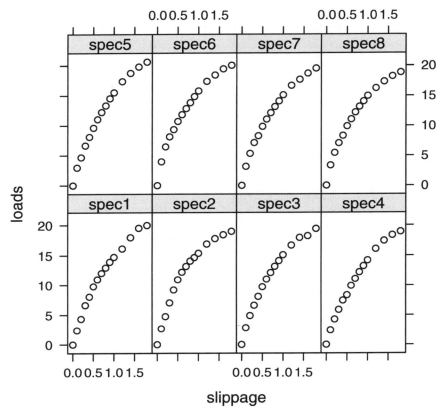

Fig. 8.1. Plot of the eight specimens in timber data.

Similar to a linear model fitted using `lm()`, the `lme()` function takes a model formula describing the response and exploratory variables, here associated with fixed effects. In addition, a second formula is needed (specified via the `random` argument) that specifies the random effects. The first model assigns a random intercept to each level of `specimen`, while the second model fits a random intercept and slope parameter in variable `slippage` for each level of `specimen`.

First we can test the random intercept model against the simple linear regression model via a likelihood ratio test for zero variances of the random-effect parameters, which is applied by the `exactLRT()` function from **RLRsim** (Scheipl 2010, Scheipl, Greven, and Küchenhoff 2008):

```
R> library("RLRsim")
R> exactRLRT(timber.lme)
```

Using restricted likelihood evaluated at ML estimators.
 Refit with method="REML" for exact results.

        simulated finite sample distribution of RLRT. (p-value
        based on 10000 simulated values)

data:
RLRT = 0, p-value = 0.4419

We see that the random intercept model does not improve upon the simple
linear model. Nevertheless, we continue to work with the mixed-effects models
simply because the "observations", due to their repeated measurement structure, cannot be assumed to be independent.
    Now we can do the same for the two random-effect models

```
R> anova(timber.lme, timber.lme1)
```

```
             Model df   AIC    BIC  logLik    Test     L.Ratio
timber.lme       1  4 466.7  477.8 -229.3
timber.lme1      2  6 470.7  487.4 -229.3 1 vs 2 1.497e-08
             p-value
timber.lme
timber.lme1      1
```

The p-value associated with the likelihood ratio test is very small, indicating
that the random intercept and slope model is to be preferred over the simpler
random intercept model for these data. The results from this model are:

```
R> summary(timber.lme1)
```

```
Linear mixed-effects model fit by maximum likelihood
 Data: timber
    AIC   BIC  logLik
  470.7 487.4 -229.3

Random effects:
 Formula: ~slippage | specimen
 Structure: General positive-definite, Log-Cholesky param.
            StdDev    Corr
(Intercept) 3.106e-04 (Intr)
slippage    1.028e-05 0
Residual    1.636e+00

Fixed effects: loads ~ slippage
                Value Std.Error  DF t-value p-value
```

```
(Intercept)   3.516    0.2644 111    13.30        0
slippage     10.373    0.2835 111    36.59        0
 Correlation:
         (Intr)
slippage -0.822
```

```
Standardized Within-Group Residuals:
    Min       Q1     Med      Q3      Max
-2.1492 -0.5997  0.2491  0.7936  1.5226
```

```
Number of Observations: 120
Number of Groups: 8
```

(Determining the degrees of freedom for the t-value given in this output is not always easy except in special cases where the data are balanced and the model for the mean has a relatively simple form; consequently, an approximation is described in Kenward and Roger 1997) The regression coefficient for slippage is highly significant. We can find the predicted values under this model and then plot them alongside a plot of the raw data using the following R code:

```
R> timber$pred1 <- predict(timber.lme1)
```

The resulting plot is shown in Figure 8.2. Clearly the fit is not good. In fact, under the random intercept and slope model, the predicted values for each specimen are almost identical, reflecting the fact that the estimated variances of both random effects are essentially zero. The plot of the observed values in Figure 8.2 shows a leveling-off of the increase in load needed to achieve a given slippage level as slippage increases; i.e., it suggests that a quadratic term in slippage is essential in any model for these data. Including this as a fixed effect, the required model is

$$y_{ij} = (\beta_0 + u_{i1}) + (\beta_1 + u_{i2})x_j + \beta_2 x_j^2 + \epsilon_{ij}. \tag{8.4}$$

The necessary R code to fit this model and test it against the previous random intercept and slope model is

```
R> timber.lme2 <- lme(loads ~ slippage + I(slippage^2),
+                     random = ~slippage | specimen,
+                     data = timber, method = "ML")
R> anova(timber.lme1, timber.lme2)
```

```
            Model df  AIC   BIC   logLik   Test L.Ratio p-value
timber.lme1     1  6 470.7 487.4 -229.33
timber.lme2     2  7 213.5 233.1  -99.77 1 vs 2   259.1  <.0001
```

The p-value from the likelihood ratio test is less than 0.0001, indicating that the model that includes a quadratic term does provide a much improved fit. Both the linear and quadratic effects of slippage are highly significant.

```
R> pfun <- function(x, y) {
+       panel.xyplot(x, y[1:length(x)])
+       panel.lines(x, y[1:length(x) + length(x)], lty = 1)
+ }
R> plot(xyplot(cbind(loads, pred1) ~ slippage | specimen,
+              data = timber, panel = pfun, layout = c(4, 2),
+              ylab = "loads"))
```

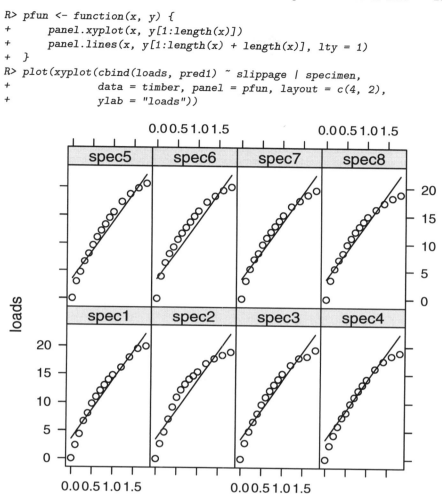

Fig. 8.2. Plot of the predicted values of the random intercept model for the timber data.

We can now produce a plot similar to that in Figure 8.2 but showing the predicted values from the model in (8.4). The code is similar to that given above and so is not repeated again here. The resulting plot is shown in Figure 8.3. Clearly the model describes the data more satisfactorily, although there remains an obvious problem, which is taken up in Exercise 8.1.

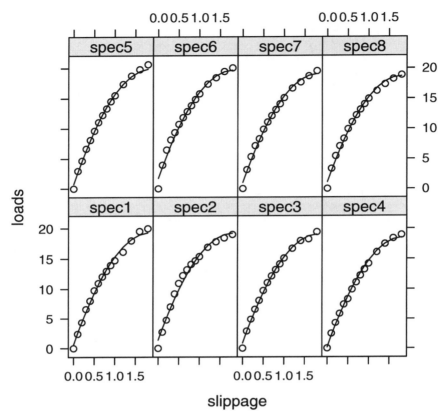

Fig. 8.3. Plot of the predicted values of the random intercept model including a quadratic term for the timber data.

8.2.3 Fitting random-effect models to the glucose challenge data

Now we can move on to consider the glucose challenge data given in its long form in Table 8.2. Again we will begin by plotting the data so that we get some ideas as to what form of linear mixed-effect model might be appropriate. First we plot the raw data separately for the control and the obese groups in Figure 8.4. First, we transform the `plasma` data from long into wide form and apply the `parallel()` function from package **lattice** to set up a parallel-coordinates plot.

The profiles in both groups show some curvature, suggesting that a quadratic effect of time may be needed in any model. There also appears to be some difference in the shapes of the curves in the two groups, suggesting perhaps the need to consider a group × time interaction. Next we plot the scatterplot matrices of the repeated measurements for the two groups using the code in Figure 8.5.

```
R> x <- reshape(plasma, direction = "wide", timevar = "time",
+               idvar = "Subject", v.names = "plasma")
R> parallel(~ x[,-(1:2)] | group, data = x, horizontal = FALSE,
+           col = "black", scales = list(x = list(labels = 1:8)),
+           ylab = "Plasma inorganic phosphate",
+           xlab = "Time (hours after oral glucose challenge)")
```

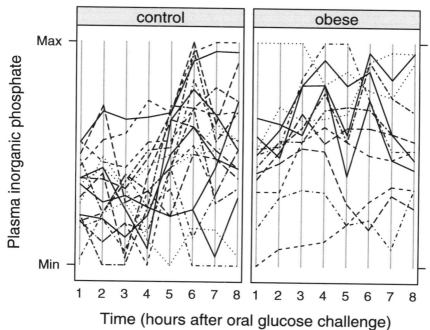

Fig. 8.4. Glucose challenge data for control and obese groups over time. Each line represents the trajectory of one individual.

The plot indicates that the correlations of pairs of measurements made at different times differ so that the compound symmetry structure for these correlations is unlikely to be appropriate.

On the basis of these plots, we will begin by fitting the model in (8.4) with the addition, in this case, of an extra covariate, namely a dummy variable coding the group, control or obese, to which a subject belongs. We can fit the required model using

```
R> plasma.lme1 <- lme(plasma ~ time + I(time^2) + group,
+                      random = ~ time | Subject,
+                      data = plasma, method = "ML")
R> summary(plasma.lme1)
```

```
R> plot(splom(~ x[, grep("plasma", colnames(x))] | group, data = x,
+              cex = 1.5, pch = ".", pscales = NULL, varnames = 1:8))
```

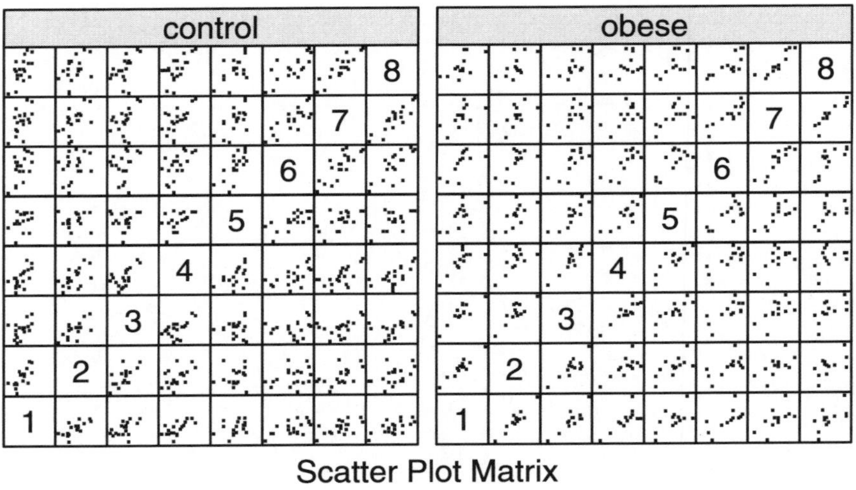

Fig. 8.5. Scatterplot matrix for glucose challenge data.

```
Linear mixed-effects model fit by maximum likelihood
 Data: plasma
    AIC   BIC logLik
  390.5 419.1 -187.2

Random effects:
 Formula: ~time | Subject
 Structure: General positive-definite, Log-Cholesky param.
            StdDev  Corr
(Intercept) 0.69772 (Intr)
time        0.09383 -0.7
Residual    0.38480

Fixed effects: plasma ~ time + I(time^2) + group
             Value Std.Error  DF  t-value p-value
(Intercept)  4.880  0.17091  229  28.552  0.0000
time        -0.803  0.05075  229 -15.827  0.0000
I(time^2)    0.085  0.00521  229  16.258  0.0000
groupobese   0.437  0.18589   31   2.351  0.0253
 Correlation:
```

```
             (Intr) time    I(t^2)
time         -0.641
I(time^2)     0.457 -0.923
groupobese   -0.428  0.000  0.000
```

```
Standardized Within-Group Residuals:
      Min         Q1        Med         Q3        Max
-2.771508  -0.548688  -0.002765   0.564435   2.889633
```

```
Number of Observations: 264
Number of Groups: 33
```

The regression coefficients for linear and quadratic time are both highly significant. The group effect is also significant, and an asymptotic 95% confidence interval for the group effect is obtained from $0.437 \pm 1.96 \times 0.186$, giving $[-3.209, 4.083]$.

Here, to demonstrate what happens if we make a very misleading assumption about the correlational structure of the repeated measurements, we will compare the results with those obtained if we assume that the repeated measurements are independent. The independence model can be fitted in the usual way with the lm() function

```
R> summary(lm(plasma ~ time + I(time^2) + group, data = plasma))
```

```
Call:
lm(formula = plasma ~ time + I(time^2) + group,
   data = plasma)
```

```
Residuals:
    Min      1Q  Median      3Q     Max
-1.6323 -0.4401  0.0347  0.4750  2.0170
```

```
Coefficients:
            Estimate Std. Error t value Pr(>|t|)
(Intercept)  4.85761    0.16686   29.11  < 2e-16
time        -0.80328    0.08335   -9.64  < 2e-16
I(time^2)    0.08467    0.00904    9.37  < 2e-16
groupobese   0.49332    0.08479    5.82  1.7e-08
```

```
Residual standard error: 0.673 on 260 degrees of freedom
Multiple R-squared: 0.328,         Adjusted R-squared: 0.32
F-statistic: 42.3 on 3 and 260 DF,  p-value: <2e-16
```

We see that, under the independence assumption, the standard error for the group effect is about one-half of that given for model plasma.lme1 and if used would lead to a much stronger claim of evidence of a difference between control and obese patients.

We will now plot the predicted values from the fitted linear mixed-effects model for each group using the code presented with Figure 8.5.

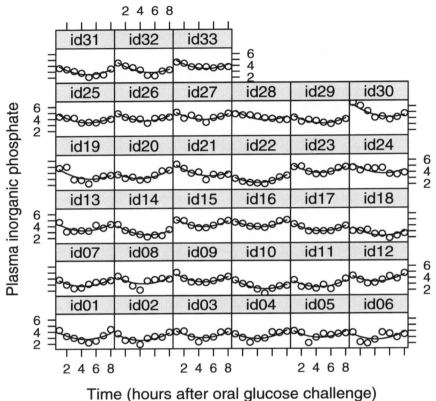

Fig. 8.6. Predictions for glucose challenge data.

We can see that the model has captured the profiles of the control group relatively well but not those of the obese group. We need to consider a further model that contains a group × time interaction.

```
R> plasma.lme2 <- lme(plasma ~ time*group +I(time^2),
+                     random = ~time | Subject,
+                     data = plasma, method = "ML")
```

The required model can be fitted and tested against the previous model using

```
R> anova(plasma.lme1, plasma.lme2)
```

```
                Model df    AIC   BIC logLik    Test L.Ratio p-value
plasma.lme1       1  8 390.5 419.1 -187.2
plasma.lme2       2  9 383.3 415.5 -182.7 1 vs 2   9.157  0.0025
```

The *p*-value associated with the likelihood ratio test is 0.0011, indicating that the model containing the interaction term is to be preferred. The results for this model are

R> summary(plasma.lme2)

```
Linear mixed-effects model fit by maximum likelihood
 Data: plasma
   AIC   BIC logLik
 383.3 415.5 -182.7
```

Random effects:
 Formula: ~time | Subject
 Structure: General positive-definite, Log-Cholesky param.
             StdDev   Corr
(Intercept) 0.64190 (Intr)
time   0.07626 -0.631
Residual   0.38480

Fixed effects: plasma ~ time * group + I(time^2)

	Value	Std.Error	DF	t-value	p-value
(Intercept)	4.659	0.17806	228	26.167	0.0000
time	-0.759	0.05178	228	-14.662	0.0000
groupobese	0.997	0.25483	31	3.911	0.0005
I(time^2)	0.085	0.00522	228	16.227	0.0000
time:groupobese	-0.112	0.03476	228	-3.218	0.0015

Correlation:

	(Intr)	time	gropbs	I(t^2)
time	-0.657			
groupobese	-0.564	0.181		
I(time^2)	0.440	-0.907	0.000	
time:groupobese	0.385	-0.264	-0.683	0.000

Standardized Within-Group Residuals:

Min	Q1	Med	Q3	Max
-2.72436	-0.53605	-0.01071	0.58568	2.95029

Number of Observations: 264
Number of Groups: 33

The interaction effect is highly significant. The fitted values from this model are shown in Figure 8.7 (the code is very similar to that given for producing Figure 8.6). The plot shows that the new model has produced predicted values that more accurately reflect the raw data plotted in Figure 8.4. The predicted profiles for the obese group are "flatter" as required.

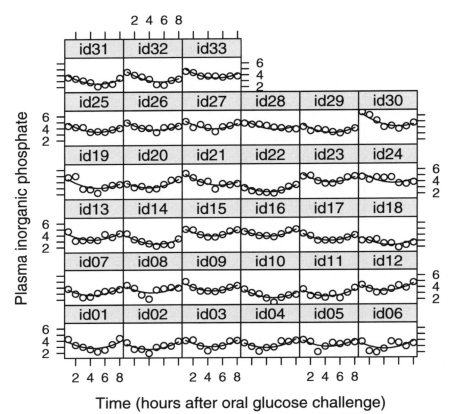

Fig. 8.7. Predictions for glucose challenge data.

We can check the assumptions of the final model fitted to the glucose challenge data (i.e., the normality of the random-effect terms and the residuals) by first using the `random.effects()` function to *predict* the former and the `resid()` function to calculate the differences between the observed data values and the fitted values and then using normal probability plots on each. How the random effects are predicted is explained briefly in Section 8.3. The necessary R code to obtain the effects, residuals, and plots is as follows:

```
R> res.int <- random.effects(plasma.lme2)[,1]
R> res.slope <- random.effects(plasma.lme2)[,2]
```

The resulting plot is shown in Figure 8.8. The plot of the residuals is linear as required, but there is some slight deviation from linearity for each of the predicted random effects.

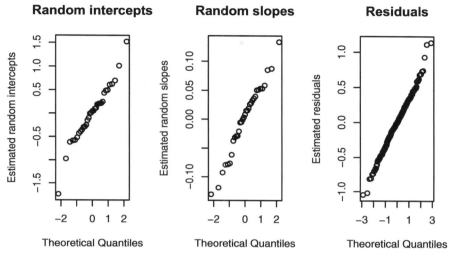

Fig. 8.8. Probability plots of predicted random intercepts, random slopes, and residuals for the final model fitted to glucose challenge data.

8.3 Prediction of random effects

The random effects are not estimated as part of the model. However, having estimated the model, we can *predict* the values of the random effects. According to Bayes' Theorem, the *posterior probability* of the random effects is given by

$$Pr(\mathbf{u}|\mathbf{y}, \mathbf{x}) = f(\mathbf{y}|\mathbf{u}, \mathbf{x})g(\mathbf{u}),$$

where $f(\mathbf{y}|\mathbf{u}, \mathbf{x})$ is the conditional density of the responses given the random effects and covariates (a product of normal densities) and $g(\mathbf{u})$ is the *prior* density of the random effects (multivariate normal). The means of this posterior distribution can be used as estimates of the random effects and are known as *empirical Bayes estimates*. The empirical Bayes estimator is also known as a shrinkage estimator because the predicted random effects are smaller in absolute value than their fixed-effect counterparts. *Best linear unbiased predictions* (BLUPs) are linear combinations of the responses that are unbiased estimators of the random effects and minimise the mean square error.

8.4 Dropouts in longitudinal data

A problem that frequently occurs when collecting longitudinal data is that some of the intended measurements are, for one reason or another, not made. In clinical trials, for example, some patients may miss one or more protocol scheduled visits after treatment has begun and so fail to have the required outcome measure taken. There will be other patients who do not complete the intended follow-up for some reason and drop out of the study before the end date specified in the protocol. Both situations result in missing values of the outcome measure. In the first case, these are intermittent, but dropping out of the study implies that once an observation at a particular time point is missing, so are all the remaining planned observations. Many studies will contain missing values of both types, although in practise it is dropouts that cause the most problems when analysing the resulting data set.

An example of a set of longitudinal data in which a number of patients have dropped out is given in Table 8.3. These data are essentially a subset of those collected in a clinical trial that is described in detail in Proudfoot, Goldberg, Mann, Everitt, Marks, and Gray (2003). The trial was designed to assess the effectiveness of an interactive program using multimedia techniques for the delivery of cognitive behavioural therapy for depressed patients and known as Beat the Blues (BtB). In a randomised controlled trial of the program, patients with depression recruited in primary care were randomised to either the BtB program or to Treatment as Usual (TAU). The outcome measure used in the trial was the Beck Depression Inventory II (see Beck, Steer, and Brown 1996), with higher values indicating more depression. Measurements of this variable were made on five occasions, one prior to the start of treatment and at two monthly intervals after treatment began. In addition, whether or not a participant in the trial was already taking anti-depressant medication was noted along with the length of time they had been depressed.

Table 8.3: `BtheB` data. Data of a randomised trial evaluating the effects of Beat the Blues.

drug	length	treatment	bdi.pre	bdi.2m	bdi.3m	bdi.5m	bdi.8m
No	>6m	TAU	29	2	2	NA	NA
Yes	>6m	BtheB	32	16	24	17	20
Yes	<6m	TAU	25	20	NA	NA	NA
No	>6m	BtheB	21	17	16	10	9
Yes	>6m	BtheB	26	23	NA	NA	NA
Yes	<6m	BtheB	7	0	0	0	0
Yes	<6m	TAU	17	7	7	3	7
No	>6m	TAU	20	20	21	19	13
Yes	<6m	BtheB	18	13	14	20	11
Yes	>6m	BtheB	20	5	5	8	12
No	>6m	TAU	30	32	24	12	2

Table 8.3: BtheB data (continued).

drug	length	treatment	bdi.pre	bdi.2m	bdi.3m	bdi.5m	bdi.8m
Yes	<6m	BtheB	49	35	NA	NA	NA
No	>6m	TAU	26	27	23	NA	NA
Yes	>6m	TAU	30	26	36	27	22
Yes	>6m	BtheB	23	13	13	12	23
No	<6m	TAU	16	13	3	2	0
No	>6m	BtheB	30	30	29	NA	NA
No	<6m	BtheB	13	8	8	7	6
No	>6m	TAU	37	30	33	31	22
Yes	<6m	BtheB	35	12	10	8	10
No	>6m	BtheB	21	6	NA	NA	NA
No	<6m	TAU	26	17	17	20	12
No	>6m	TAU	29	22	10	NA	NA
No	>6m	TAU	20	21	NA	NA	NA
No	>6m	TAU	33	23	NA	NA	NA
No	>6m	BtheB	19	12	13	NA	NA
Yes	<6m	TAU	12	15	NA	NA	NA
Yes	>6m	TAU	47	36	49	34	NA
Yes	>6m	BtheB	36	6	0	0	2
No	<6m	BtheB	10	8	6	3	3
No	<6m	TAU	27	7	15	16	0
No	<6m	BtheB	18	10	10	6	8
Yes	<6m	BtheB	11	8	3	2	15
Yes	<6m	BtheB	6	7	NA	NA	NA
Yes	>6m	BtheB	44	24	20	29	14
No	<6m	TAU	38	38	NA	NA	NA
No	<6m	TAU	21	14	20	1	8
Yes	>6m	TAU	34	17	8	9	13
Yes	<6m	BtheB	9	7	1	NA	NA
Yes	>6m	TAU	38	27	19	20	30
Yes	<6m	BtheB	46	40	NA	NA	NA
No	<6m	TAU	20	19	18	19	18
Yes	>6m	TAU	17	29	2	0	0
No	>6m	BtheB	18	20	NA	NA	NA
Yes	>6m	BtheB	42	1	8	10	6
No	<6m	BtheB	30	30	NA	NA	NA
Yes	<6m	BtheB	33	27	16	30	15
No	<6m	BtheB	12	1	0	0	NA
Yes	<6m	BtheB	2	5	NA	NA	NA
No	>6m	TAU	36	42	49	47	40
No	<6m	TAU	35	30	NA	NA	NA
No	<6m	BtheB	23	20	NA	NA	NA
No	>6m	TAU	31	48	38	38	37

Table 8.3: BtheB data (continued).

drug	length	treatment	bdi.pre	bdi.2m	bdi.3m	bdi.5m	bdi.8m
Yes	<6m	BtheB	8	5	7	NA	NA
Yes	<6m	TAU	23	21	26	NA	NA
Yes	<6m	BtheB	7	7	5	4	0
No	<6m	TAU	14	13	14	NA	NA
No	<6m	TAU	40	36	33	NA	NA
Yes	<6m	BtheB	23	30	NA	NA	NA
No	>6m	BtheB	14	3	NA	NA	NA
No	>6m	TAU	22	20	16	24	16
No	>6m	TAU	23	23	15	25	17
No	<6m	TAU	15	7	13	13	NA
No	>6m	TAU	8	12	11	26	NA
No	>6m	BtheB	12	18	NA	NA	NA
No	>6m	TAU	7	6	2	1	NA
Yes	<6m	TAU	17	9	3	1	0
Yes	<6m	BtheB	33	18	16	NA	NA
No	<6m	TAU	27	20	NA	NA	NA
No	<6m	BtheB	27	30	NA	NA	NA
No	<6m	BtheB	9	6	10	1	0
No	>6m	BtheB	40	30	12	NA	NA
No	>6m	TAU	11	8	7	NA	NA
No	<6m	TAU	9	8	NA	NA	NA
No	>6m	TAU	14	22	21	24	19
Yes	>6m	BtheB	28	9	20	18	13
No	>6m	BtheB	15	9	13	14	10
Yes	>6m	BtheB	22	10	5	5	12
No	<6m	TAU	23	9	NA	NA	NA
No	>6m	TAU	21	22	24	23	22
No	>6m	TAU	27	31	28	22	14
Yes	>6m	BtheB	14	15	NA	NA	NA
No	>6m	TAU	10	13	12	8	20
Yes	<6m	TAU	21	9	6	7	1
Yes	>6m	BtheB	46	36	53	NA	NA
No	>6m	BtheB	36	14	7	15	15
Yes	>6m	BtheB	23	17	NA	NA	NA
Yes	>6m	TAU	35	0	6	0	1
Yes	<6m	BtheB	33	13	13	10	8
No	<6m	BtheB	19	4	27	1	2
No	<6m	TAU	16	NA	NA	NA	NA
Yes	<6m	BtheB	30	26	28	NA	NA
Yes	<6m	BtheB	17	8	7	12	NA
No	>6m	BtheB	19	4	3	3	3
No	>6m	BtheB	16	11	4	2	3

Table 8.3: BtheB data (continued).

drug	length	treatment	bdi.pre	bdi.2m	bdi.3m	bdi.5m	bdi.8m
Yes	>6m	BtheB	16	16	10	10	8
Yes	<6m	TAU	28	NA	NA	NA	NA
No	>6m	BtheB	11	22	9	11	11
No	<6m	TAU	13	5	5	0	6
Yes	<6m	TAU	43	NA	NA	NA	NA

To begin, we shall graph the data here by plotting the boxplots of each of the five repeated measures separately for each treatment group. Assuming the data are available as the data frame `BtheB`, the necessary code is given with Figure 8.9.

Figure 8.9 shows that there is a decline in BDI values in both groups, with perhaps the values in the BtheB group being lower at each post-randomisation visit. We shall fit both random intercept and random intercept and slope models to the data including the pre-BDI values, treatment group, drugs, and length as fixed-effect covariates. First we need to rearrange the data into the long form using the following code:

```
R> BtheB$subject <- factor(rownames(BtheB))
R> nobs <- nrow(BtheB)
R> BtheB_long <- reshape(BtheB, idvar = "subject",
+        varying = c("bdi.2m", "bdi.3m", "bdi.5m", "bdi.8m"),
+        direction = "long")
R> BtheB_long$time <- rep(c(2, 3, 5, 8), rep(nobs, 4))
```

The resulting data frame `BtheB_long` contains a number of missing values, and in applying the `lme()` function these will need to be dropped. But notice it is only the missing values that are removed, *not* participants that have at least one missing value. All the available data are used in the model-fitting process. We can fit the two models and test which is most appropriate using

```
R> BtheB_lme1 <- lme(bdi ~ bdi.pre + time + treatment + drug +
+        length, random = ~ 1 | subject, data = BtheB_long,
+        na.action = na.omit)
R> BtheB_lme2 <- lme(bdi ~ bdi.pre + time + treatment + drug +
+        length, random = ~ time | subject, data = BtheB_long,
+        na.action = na.omit)
```

This results in

```
R> anova(BtheB_lme1, BtheB_lme2)

            Model df  AIC  BIC  logLik   Test L.Ratio p-value
BtheB_lme1      1  8 1883 1912 -933.5
BtheB_lme2      2 10 1886 1922 -933.2 1 vs 2  0.5665  0.7533
```

```
R> ylim <- range(BtheB[,grep("bdi", names(BtheB))],
+                   na.rm = TRUE)
R> tau <- subset(BtheB, treatment == "TAU")[,
+      grep("bdi", names(BtheB))]
R> boxplot(tau, main = "Treated as Usual", ylab = "BDI",
+         xlab = "Time (in months)", names = c(0, 2, 3, 5, 8),
+         ylim = ylim)
R> btheb <- subset(BtheB, treatment == "BtheB")[,
+      grep("bdi", names(BtheB))]
R> boxplot(btheb, main = "Beat the Blues", ylab = "BDI",
+         xlab = "Time (in months)", names = c(0, 2, 3, 5, 8),
+         ylim = ylim)
```

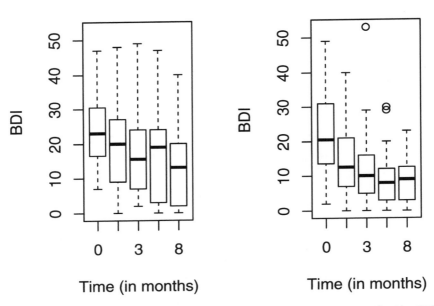

Fig. 8.9. Boxplots for the repeated measures by treatment group for the BtheB data.

Clearly, the simpler random intercept model is adequate for these data. The results from this model can be found using

```
R> summary(BtheB_lme1)
```

```
Linear mixed-effects model fit by REML
 Data: BtheB_long
   AIC  BIC logLik
   1883 1912 -933.5
```

```
Random effects:
 Formula: ~1 | subject
         (Intercept) Residual
StdDev:       7.206    5.029

Fixed effects: bdi ~ bdi.pre + time + treatment + drug + length
                Value Std.Error  DF t-value p-value
(Intercept)     5.574    2.2995 182   2.424  0.0163
bdi.pre         0.640    0.0799  92   8.013  0.0000
time           -0.702    0.1469 182  -4.775  0.0000
treatmentBtheB -2.315    1.7152  92  -1.350  0.1804
drugYes        -2.816    1.7729  92  -1.588  0.1156
length>6m       0.179    1.6816  92   0.106  0.9155
 Correlation:
                (Intr) bdi.pr time   trtmBB drugYs
bdi.pre        -0.683
time           -0.232  0.019
treatmentBtheB -0.390  0.121  0.017
drugYes        -0.074 -0.236 -0.022 -0.323
length>6m      -0.244 -0.241 -0.036  0.002  0.157

Standardized Within-Group Residuals:
     Min        Q1        Med        Q3        Max
-2.68699  -0.50847  -0.06085  0.42067  3.81414

Number of Observations: 280
Number of Groups: 97
```

The effect of most interest in this study is, of course, the treatment effect, and our analysis shows that this is not significant at the 5% level. The only effect that is significant is time, the negative value of its regression coefficient showing that the BDI values decline over the eight months of the study.

We now need to consider briefly how the dropouts may affect the analyses reported above. To understand the problems that patients dropping out can cause for the analysis of data from a longitudinal trial, we need to consider a classification of dropout mechanisms first introduced by Rubin (1976). The type of mechanism involved has implications for which approaches to analysis are suitable and which are not. Rubin's suggested classification involves three types of dropout mechanisms:

- *Dropout completely at random* (DCAR). Here the probability that a patient drops out does not depend on either the observed or missing values of the response. Consequently, the observed (non-missing) values effectively constitute a simple random sample of the values for all subjects. Possible examples include missing laboratory measurements because of a dropped test tube (if it was not dropped because of the knowledge of any measurement), the accidental death of a participant in a study, or a participant moving to another area. Intermittent missing values in a longitudinal data set, whereby a patient misses a clinic visit for transitory reasons ("went shopping instead" or the like) can reasonably be assumed to be DCAR. Completely random dropout causes the least problems for data analysis, but it is a strong assumption.

- *Dropout at random* (DAR). The dropout at random mechanism occurs when the probability of dropping out depends on the outcome measures that have been observed in the past but given this information is conditionally independent of all the future (unrecorded) values of the outcome variable following dropout. Here "missingness" depends only on the observed data, with the distribution of future values for a subject who drops out at a particular time being the same as the distribution of the future values of a subject who remains in at that time, if they have the same covariates and the same past history of outcome up to and including the specific time point. Murray and Findlay (1988) provide an example of this type of missing value from a study of hypertensive drugs in which the outcome measure was diastolic blood pressure. The protocol of the study specified that the participant was to be removed from the study when his or her blood pressure got too large. Here blood pressure at the time of dropout was observed before the participant dropped out, so although the dropout mechanism is not DCAR since it depends on the values of blood pressure, it *is* DAR, because dropout depends only on the observed part of the data. A further example of a DAR mechanism is provided by Heitjan (1997) and involves a study in which the response measure is body mass index (BMI). Suppose that the measure is missing because subjects who had high body mass index values at earlier visits avoided being measured at later visits out of embarrassment, regardless of whether they had gained or lost weight in the intervening period. The missing values here are DAR but *not* DCAR; consequently methods applied to the data that assumed the latter might give misleading results (see later discussion).

- *Non-ignorable* (sometimes referred to as *informative*). The final type of dropout mechanism is one where the probability of dropping out depends on the unrecorded missing values–observations are likely to be missing when the outcome values that would have been observed had the patient not dropped out are systematically higher or lower than usual (corresponding perhaps to their condition becoming worse or improving). A non-medical example is when individuals with lower income levels or very high incomes are less likely to provide their personal income in an interview. In a medical setting, possible examples are a participant dropping out of a longitudinal study when his or her blood pressure became too high and this value was not observed, or when their pain became intolerable and we did not record the associated pain value. For the BDI example introduced above, if subjects were more likely to avoid being measured if they had put on extra weight since the last visit, then the data are non-ignorably missing. Dealing with data containing missing values that result from this type of dropout mechanism is difficult. The correct analyses for such data must estimate the dependence of the missingness probability on the missing values. Models and software that attempt this are available (see for example, Diggle and Kenward 1994), but their use is not routine and, in addition, it must be remembered that the associated parameter estimates can be unreliable.

Under what type of dropout mechanism are the mixed-effects models considered in this chapter valid? The good news is that such models can be shown to give valid results under the relatively weak assumption that the dropout mechanism is DAR (Carpenter, Pocock, and Lamm 2002). When the missing values are thought to be informative, any analysis is potentially problematical. But Diggle and Kenward (1994) have developed a modelling framework for longitudinal data with informative dropouts, in which random or completely random dropout mechanisms are also included as explicit models. The essential feature of the procedure is a logistic regression model for the probability of dropping out, in which the explanatory variables can include previous values of the response variable and, in addition, the *unobserved* value at dropout as a *latent* variable (i.e., an unobserved variable). In other words, the dropout probability is allowed to depend on both the *observed* measurement history and the unobserved value at dropout. This allows both a formal assessment of the type of dropout mechanism in the data and the estimation of effects of interest, for example, treatment effects under different assumptions about the dropout mechanism. A full technical account of the model is given in Diggle and Kenward (1994), and a detailed example that uses the approach is described in Carpenter et al. (2002).

One of the problems for an investigator struggling to identify the dropout mechanism in a data set is that there are no routine methods to help, although a number of largely ad hoc graphical procedures can be used as described in Diggle (1998), Everitt (2002), and Carpenter et al. (2002).

We shall now illustrate one very simple graphical procedure for assessing the dropout mechanism suggested in Carpenter et al. (2002). That involves plotting the observations for each treatment group, at each time point, differentiating between two categories of patients: those who do and those who do not attend their next scheduled visit. Any clear difference between the distributions of values for these two categories indicates that dropout is not completely at random. For the Beat the Blues data, such a plot is shown in Figure 8.10. When comparing the distribution of BDI values for patients who do (circles) and do not (bullets) attend the next scheduled visit, there is no apparent difference, and so it is reasonable to assume dropout completely at random.

```
R> bdi <- BtheB[, grep("bdi", names(BtheB))]
R> plot(1:4, rep(-0.5, 4), type = "n", axes = FALSE,
+        ylim = c(0, 50), xlab = "Months", ylab = "BDI")
R> axis(1, at = 1:4, labels = c(0, 2, 3, 5))
R> axis(2)
R> for (i in 1:4) {
+        dropout <- is.na(bdi[,i + 1])
+        points(rep(i, nrow(bdi)) + ifelse(dropout, 0.05, -0.05),
+               jitter(bdi[,i]), pch = ifelse(dropout, 20, 1))
+ }
```

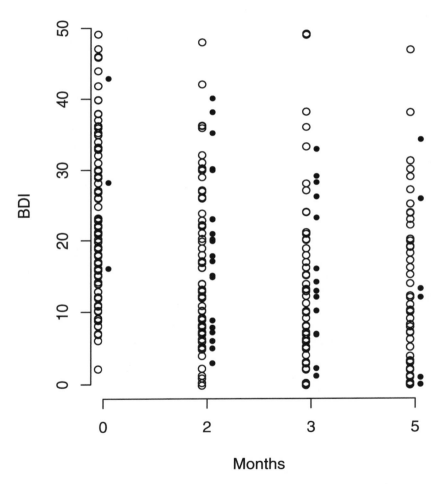

Fig. 8.10. Distribution of BDI values for patients who do (circles) and do not (bullets) attend the next scheduled visit.

8.5 Summary

Linear mixed-effects models are extremely useful for modelling longitudinal data in particular and repeated measures data more generally. The models allow the correlations between the repeated measurements to be accounted for so that correct inferences can be drawn about the effects of covariates of interest on the repeated response values. In this chapter, we have concentrated on responses that are continuous and conditional on the explanatory variables, and random effects have a normal distribution. But random-effects models can also be applied to non-normal responses, for example binary variables–see, for example Everitt (2002) and Skrondal and Rabe-Hesketh (2004).

The lack of independence of repeated measures data is what makes the modelling of such data a challenge. But even when only a single measurement of a response is involved, correlation can, in some circumstances, occur between the response values of different individuals and cause similar problems. As an example, consider a randomised clinical trial in which subjects are recruited at multiple study centres. The multicentre design can help to provide adequate sample sizes and enhance the generalisability of the results. However, factors that vary by centre, including patient characteristics and medical practise patterns, may exert a sufficiently powerful effect to make inferences that ignore the "clustering" seriously misleading. Consequently, it may be necessary to incorporate random effects for centres into the analysis.

8.6 Exercises

Ex. 8.1 The final model fitted to the timber data did not constrain the fitted curves to go through the origin, although this is clearly necessary. Fit an amended model where this constraint is satisfied, and plot the new predicted values.

Ex. 8.2 Investigate a further model for the glucose challenge data that allows a random quadratic effect.

Ex. 8.3 Fit an independence model to the Beat the Blues data, and compare the estimated treatment effect confidence interval with that from the random intercept model described in the text.

Ex. 8.4 Construct a plot of the mean profiles of the two treatment groups in the Beat the Blues study showing also the predicted mean profiles under the model used in the chapter. Repeat the exercise with a model that includes only a time effect.

Ex. 8.5 Investigate whether there is any evidence of an interaction between treatment and time for the Beat the Blues data.

References

Affifi, A. A., Clark, V. A., and May, S. (2004), *Computer-Aided Multivariate Analysis*, London, UK: Chapman & Hall, 4th edition. Cited on p. 99.

Anonymous (1988), "Statistical abstract of the USA," Technical Report 159, U.S. Department of Commerce, Washington, DC. Cited on p. 176.

Banfield, J. D. and Raftery, A. E. (1993), "Model-based Gaussian and non-Gaussian clustering," *Biometrics*, 49, 803–821. Cited on p. 186, 188.

Barlow, R. E., Bartholomew, D. J., Bremner, J. M., and Brunk, H. D. (1972), *Statistical Inference under Order Restrictions*, New York: John Wiley & Sons. Cited on p. 123.

Bartholomew, D. and Knott, M. (1987), *Latent Variable Models and Factor Analysis*, London, UK: Griffin. Cited on p. 147.

Bartholomew, D. J. (2005), "History of factor analysis: A statistical perspective," in *Encyclopedia of Statistics in Behavioral Science*, eds. B. S. Everitt and D. Howell, Chichester, UK: John Wiley & Sons. Cited on p. 136.

Bartlett, M. S. (1947), "Multivariate analysis," *Journal of the Royal Statistical Society, Series B*, 9, 176–197. Cited on p. 103.

Beck, A., Steer, R., and Brown, G. (1996), BDI-II Manual, The Psychological Corporation, San Antonio, TX, 2nd edition. Cited on p. 248.

Becker, C. and Gather, U. (2001), "The largest nonidentifiable outlier: A comparison of multivariate simultaneous outlier identification rules," *Computational Statistics & Data Analysis*, 36, 119–127. Cited on p. 23.

Becker, R., Cleveland, W., Shyu, M., and Kaluzny, S. (1994), "Trellis displays: User's guide," Technical report, AT&T Research, URL http://www2.research.att.com/areas/stat/doc/94.10.ps. Cited on p. 50, 57.

Bellman, R. (1961), *Adaptive Control Processes*, Princeton, NJ: Princeton University Press. Cited on p. 61.

Bentler, P. (1980), "Multivariate analysis with latent variables: Causal modeling," *Annual Review of Psychology*, 31, 419–456. Cited on p. 202.

Bentler, P. (1982), "Linear systems with multiple levels and types of latent variables," in *Systems under Indirect Observation*, eds. K. G. Jöreskog and H. Wold, Amsterdam: North-Holland, pp. 101–130. Cited on p. 222.

Benzécri, J. (1992), *Correspondence Analysis Handbook*, New York: Marcel Dekker Inc. Cited on p. 127.

Blackith, R. E. and Reyment, R. A. (1971), *Multivariate Morphometrics*, London, UK: Academic Press. Cited on p. 144.

Blalock, H. M. J. (1961), "Correlation and causality: The multivariate case," *Social Forces*, 39, 246–251. Cited on p. 202.

Blalock, H. M. J. (1963), "Making causal inferences for unmeasured variables from correlations among indicators," *The American Journal of Sociology*, 69, 53–62. Cited on p. 202.

Bollen, K. and Long, J. (1993), *Testing Structural Equation Models*, Newbury Park, CA: Sage Publications. Cited on p. 204, 205.

Bouguila, N. and Amayri, O. (2009), "A discrete mixture-based kernel for SVMs: Application to spam and image categorization," *Information Processing and Management*, 45, 631–642. Cited on p. 186.

Branchaud, E. A., Cham, J. G., Nenadic, Z., Andersen, R. A., and Burdick, J. (2010), "A miniature robot for autonomous single neuron recordings," *Proceedings of the 2005 IEEE International Conference on Robotics and Automation*, pp. 1920–1926. Cited on p. 186.

Browne, M. (1982), "Covariance structures," in *Topics in Applied Multivariate Analysis*, ed. D. M. Hawkins, London, UK: Cambridge University Press. Cited on p. 204, 222.

Browne, M. W. (1974), "Generalized least squares estimates in the analysis of covariance structures," *South African Statistical Journal*, 8, 1–24. Cited on p. 202.

Calsyn, J. R. and Kenny, D. A. (1977), "Self-concept of ability and perceived evaluation of others. Cause or effect of academic achievement?" *Journal of Educational Psychology*, 69, 136–145. Cited on p. 206.

Carpenter, J., Pocock, S., and Lamm, C. J. (2002), "Coping with missing data in clinical trials: A model-based approach applied to asthma trials," *Statistics in Medicine*, 21, 1043–1066. Cited on p. 255.

Carroll, J., Clark, L., and DeSarbo, W. (1984), "The representation of three-way proximity data by single and multiple tree structure models," *Journal of Classification*, 1, 25–74. Cited on p. 106.

Carroll, J. B. (1953), "An analytical solution for approximating simple structure in factor analysis," *Psychometrika*, 18, 23–38. Cited on p. 146.

Carroll, J. D. and Arabie, P. (1980), "Multidimensional scaling," *Annual Review of Psychology*, 31, 607–649. Cited on p. 106.

Cattell, R. B. (1966), "The scree test for the number of factors," *Multivariate Behavioural Research*, 1, 245–276. Cited on p. 72.

Celeux, G. and Govaert, G. (1995), "A classification EM algorithm for clustering and two stochastic versions," *Computational Statistics & Data Analysis*, 14, 315–332. Cited on p. 188.

Chambers, J. M., Cleveland, W. S., Kleiner, B., and Tukey, P. A. (1983), *Graphical Methods for Data Analysis*, London, UK: Chapman & Hall/CRC. Cited on p. 25.

Chatfield, C. and Collins, A. (1980), *Introduction to Multivariate Analysis*, London, UK: Chapman & Hall. Cited on p. 159.

Chernoff, H. (1973), "The use of faces to represent points in k-dimensional space graphically," *Journal of the American Statistical Association*, 68, 361–368. Cited on p. 38.

Cliff, N. (1983), "Some cautions concerning the application of causal modeling methods," *Multivariate Behavioral Research*, 18, 115–126. Cited on p. 222.

Cook, D. and Swayne, D. F. (2007), *Interactive and Dynamic Graphics for Data Analysis*, New York: Springer-Verlag. Cited on p. 26, 199.

Corbet, G. B., Cummins, J., Hedges, S. R., and Krzanowski, W. J. (1970), "The taxonomic structure of British water voles, genus *Arvicola*," *Journal of Zoology*, 61, 301–316. Cited on p. 117.

Crowder, M. J. (1998), "Nonlinear growth curve," in *Encyclopedia of Biostatistics*, eds. P. Armitage and T. Colton, Chichester, UK: John Wiley & Sons. Cited on p. 226.

Dai, X. F., Erkkila, T., Yli-Harja, O., and Lahdesmaki, H. (2009), "A joint finite mixture model for clustering genes from independent Gaussian and beta distributed data," *BMC Bioinformatics*, 10. Cited on p. 186.

Dalgaard, P. (2002), *Introductory Statistics with R*, New York: Springer-Verlag. Cited on p. viii.

Davis, C. S. (2003), *Statistical Methods in the Analysis of Repeated Measurements*, New York: Springer-Verlag. Cited on p. 226, 232.

Day, N. (1969), "Estimating the components of a mixture of normal distributions," *Biometrika*, 56, 463–474. Cited on p. 187.

de Leeuw, J. (1983), "On the prehistory of correspondence analysis," *Statistica Neerlandica*, 37, 161–164. Cited on p. 127.

Dempster, A., Laird, N., and Rubin, D. (1977), "Maximum likelihood from incomplete data via the EM algorithm," *Journal of the Royal Statistical Society, Series B*, 39, 1–38. Cited on p. 188.

Diggle, P. J. (1998), "Dealing with missing values in longitudinal studies," in *Statistical Analysis of Medical Data*, eds. B. S. Everitt and G. Dunn, London, UK: Arnold. Cited on p. 255.

Diggle, P. J., Heagerty, P. J., Liang, K. Y., and Zeger, S. L. (2003), *Analysis of Longitudinal Data*, Oxford, UK: Oxford University Press. Cited on p. 232, 235.

Diggle, P. J. and Kenward, M. G. (1994), "Informative dropout in longitudinal data analysis," *Journal of the Royal Statistical Society, Series C*, 43, 49–93. Cited on p. 255.

Duncan, O. (1969), "Some linear models for two-wave, two-variable panel analysis," *Psychological Bulletin*, 72, 177–182. Cited on p. 202.

Dunson, D. B. (2009), "Bayesian nonparametric hierarchical modeling," *Biometrical Journal*, 51, 273–284. Cited on p. 186.

Everitt, B. (1984), *An Introduction to Latent Variable Models*, London, UK: Chapman & Hall. Cited on p. 142, 187, 203.

Everitt, B. (1987), *Introduction to Optimization Methods and Their Application in Statistics*, London, UK: Chapman & Hall. Cited on p. 142.

Everitt, B. and Bullmore, E. (1999), "Mixture model mapping of brain activation in functional magnetic resonance images," *Human Brain Mapping*, 7, 1–14. Cited on p. 186.

Everitt, B. and Hand, D. (1981), *Finite Mixture Distributions*, Chapman & Hall/CRC. Cited on p. 186, 187, 188.

Everitt, B. S. (2002), *Modern Medical Statistics*, London, UK: Arnold. Cited on p. 255, 257.

Everitt, B. S. and Dunn, G. (2001), *Applied Multivariate Data Analysis*, London, UK: Arnold, 2nd edition. Cited on p. 203, 209.

Everitt, B. S., Landau, S., Leese, M., and Stahl, D. (2011), *Cluster Analysis*, Chichester, UK: John Wiley & Sons, 5th edition. Cited on p. 165, 166, 171, 176, 180, 198.

Everitt, B. S. and Rabe-Hesketh, S. (1997), *The Analysis of Proximity Data*, London, UK: Arnold. Cited on p. 106.

Farmer, S. A. (1971), "An investigation into the results of principal components analysis of data derived from random numbers," *Statistician*, 20, 63–72. Cited on p. 72.

Fayyad, U. M., Piatetsky-Shapiro, G., Smyth, P., and Uthurusamy, R., eds. (1996), *Advances in Knowledge Discovery and Data Mining*, Menlo Park, CA: AAAI Press/The MIT Press. Cited on p. 4.

Fisher, N. I. and Switzer, P. (1985), "Chi-plots for assessing dependence," *Biometrika*, 72, 253–265. Cited on p. 34.

Fisher, N. I. and Switzer, P. (2001), "Graphical assessment of dependence: Is a picture worth a 100 tests?" *The American Statistician*, 55, 233–239. Cited on p. 34.

Fox, J., Kramer, A., and Friendly, M. (2010), sem: Structural Equation Models, URL http://CRAN.R-project.org/package=sem, R package version 0.9-21. Cited on p. 207.

Fraley, C. and Raftery, A. (2010), mclust: Model-Based Clustering / Normal Mixture Modeling, URL http://CRAN.R-project.org/package=mclust, R package version 3.4.8. Cited on p. 191.

Fraley, C. and Raftery, A. E. (2002), "Model-based clustering, discriminant analysis, and density estimation," *Journal of the American Statistical Association*, 97, 611–631. Cited on p. 188.

Fraley, C. and Raftery, A. E. (2003), "Enhanced model-based clustering, density estimation, and discriminant analysis software: MCLUST," *Journal of Classification*, 20, 263–286. Cited on p. 188.

Fraley, C. and Raftery, A. E. (2007), "Bayesian regularization for normal mixture estimation and model-based clustering," *Journal of Classification*, 24, 155–181. Cited on p. 188.

Frets, G. P. (1921), "Heredity of head form in man," *Genetica*, 3, 193–384. Cited on p. 96.

Frühwirth-Schnatter, S. (2006), *Finite Mixture and Markov Switching Models*, New York: Springer-Verlag. Cited on p. 186.

Gabriel, K. R. (1971), "The biplot graphic display of matrices with application to principal component analysis," *Biometrika*, 58, 453–467. Cited on p. 92.

Gabriel, K. R. (1981), "Biplot display of multivariate matrices for inspection of data and diagnosis," in *Interpreting Multivariate Data*, ed. V. Barnett, Chichester, UK: John Wiley & Sons. Cited on p. 93.

Ganesalingam, J., Stahl, D., Wijesekera, L., Galtrey, C., Shaw, C. E., Leigh, P. N., and Al-Chalabi, A. (2009), "Latent cluster analysis of ALS phenotypes identifies prognostically differing groups," *PLoS One*, 4. Cited on p. 186.

Gansner, E. and North, S. (2000), "An open graph visualization system and its applications to software engineering," *Software-Practice & Experience*, 30, 1203–1233. Cited on p. 210.

Goldberg, K. and Iglewicz, B. (1992), "Bivariate extensions of the boxplot," *Technometrics*, 34, 307–320. Cited on p. 28, 31.

Gordon, A. (1999), *Classification*, Boca Raton, FL: Chapman & Hall/CRC, 2nd edition. Cited on p. 165.

Gordon, A. D. (1987), "A review of hierarchical classification," *Journal of the Royal Statistical Society, Series A*, 150, 119–137. Cited on p. 165.

Gorsuch, R. L. (1983), *Factor Analysis*, Hillsdale, NJ: Lawrence Erlbaum Associates, 2nd edition. Cited on p. 93.

Gower, J. (1966), "Some distance properties of latent root and vector methods used in multivariate analysis," *Biometrika*, 53, 325–338. Cited on p. 109.

Gower, J. C. and Hand, D. J. (1996), *Biplots*, London, UK: Chapman & Hall/CRC. Cited on p. 93.

Gower, J. C. and Ross, G. J. S. (1969), "Minimum spanning trees and single linkage cluster analysis," *Applied Statistics*, 18, 54–64. Cited on p. 121.

Greenacre, M. (1992), "Correspondence analysis in medical research," *Statistical Methods in Medical Research*, 1, 97. Cited on p. 129, 131.

Greenacre, M. (2007), *Correspondence Analysis in Practice*, Boca Raton, FL: Chapman & Hall/CRC, 2nd edition. Cited on p. 128.

Guadagnoli, E. and Velicer, W. F. (1988), "Relation of sample size to the stability of component patterns," *Psychological Bulletin*, 103, 265–275. Cited on p. 93.

Hand, D. J., Daly, F., Lunn, A. D., McConway, K. J., and Ostrowski, E. (1994), *A Handbook of Small Datasets*, London, UK: Chapman & Hall/CRC. Cited on p. 79, 171.

Hand, D. J., Mannila, H., and Smyth, P. (2001), *Principles of Data Mining*, Cambridge, MA: The MIT Press. Cited on p. 4.

Hasselblad, V. (1966), "Estimation of parameters for a mixture of normal distributions," *Technometrics*, 8, 431–444. Cited on p. 187.

Hasselblad, V. (1969), "Estimation of finite mixtures of distributions from the exponential family," *Journal of the American Statistical Association*, 64, 1459–1471. Cited on p. 187.

Hatcher, L. (1994), *A Step-by-Step Approach to Using the SAS System for Factor Analysis and Structural Equation Modeling*, Cary, NC: SAS Institute. Cited on p. 93.

Heitjan, D. F. (1997), "Annotation: What can be done about missing data? Approaches to imputation," *American Journal of Public Health*, 87, 548–550. Cited on p. 254.

Hendrickson, A. and White, P. (1964), "Promax: A quick method for rotation to oblique simple structure," *British Journal of Mathematical and Statistical Psychology*, 17, 65–70. Cited on p. 147.

Hershberger, S. L. (2005), "Factor score estimation," in *Encyclopedia of Behavioral Statistics*, eds. B. S. Everitt and D. C. Howell, Chichester, UK: John Wiley & Sons. Cited on p. 148.

Heywood, H. (1931), "On finite sequences of real numbers," *Proceedings of the Royal Society of London, Series A, Containing Papers of a Mathematical and Physical Character*, 134, 486–501. Cited on p. 142.

Hills, M. (1977), "Book review," *Applied Statistics*, 26, 339–340. Cited on p. 159.

Hotelling, H. (1933), "Analysis of a complex of statistical variables into principal components," *Journal of Educational Psychology*, 24, 417–441. Cited on p. 101.

Huba, G. J., Wingard, J., and Bentler, P. (1981), "A comparison of two latent variable causal models for adolescent drug use," *Journal of Personality and Social Psychology*, 40, 180–193. Cited on p. 151, 211, 216.

Hyvärinen, A. (1999), "The fixed-plot algorithm and maximum likelihood estimation for independent component analysis," *Neural Processing Letters*, 10, 1–5. Cited on p. 102.

Hyvärinen, A., Karhunen, L., and Oja, E. (2001), *Independent Component Analysis*, New York: John Wiley & Sons. Cited on p. 102.

Izenman, A. J. (2008), *Modern Multivariate Techniques*, New York: Springer-Verlag. Cited on p. 102.

Jennrich, R. and Sampson, P. (1966), "Rotation for simple loadings," *Psychometrika*, 31, 313–323. Cited on p. 146.

Jolliffe, I. (1972), "Discarding variables in a principal component analysis. I: Artificial data," *Journal of the Royal Statistical Society, Series C*, 21, 160–173. Cited on p. 72.

Jolliffe, I. (1989), "Rotation of ill-defined principal components," *Journal of the Royal Statistical Society, Series C*, 38, 139–148. Cited on p. 147.

Jolliffe, I. (2002), *Principal Component Analysis*, New York: Springer-Verlag. Cited on p. 63, 64, 66, 71, 89, 90, 142.

Jones, M. C. and Sibson, R. (1987), "What is projection pursuit?" *Journal of the Royal Statistical Society, Series A*, 150, 1–38. Cited on p. 199.

Jöreskog, K. G. (1973), "Analysis of covariance structures," in *Multivariate Analysis, Volume II*, ed. P. R. Krishnaiah, New York: Academic Press. Cited on p. 202.

Jöreskog, K. G. and Sörbom, D. (1981), "Analysis of linear structural relationships by maximum likelihood and least squares," Technical Report 81-8, University of Uppsala, Uppsala, Sweden. Cited on p. 221.

Kaiser, H. (1958), "The varimax criterion for analytic rotation in factor analysis," *Psychometrika*, 23, 187–200. Cited on p. 72, 146.

Kaufman, L. and Rousseeuw, P. J. (1990), *Finding Groups in Data: An Introduction to Cluster Analysis*, New York: John Wiley & Sons. Cited on p. 133.

Kenward, M. G. and Roger, J. H. (1997), "Small sample inference for fixed effects from restricted maximum likelihood," *Biometrics*, 53, 983–997. Cited on p. 238.

Keyfitz, N. and Flieger, W. (1971), *Population: Facts and Methods of Demography*, San Francisco, CA: W.H. Freeman. Cited on p. 148.

Kruskal, J. B. (1964a), "Multidimensional scaling by optimizing goodness-of-fit to a nonmetric hypothesis," *Psychometrika*, 29, 1–27. Cited on p. 121.

Kruskal, J. B. (1964b), "Nonmetric multidimensional scaling: A numerical method," *Psychometrika*, 29, 115–129. Cited on p. 123.

Krzanowski, W. J. (1988), *Principles of Multivariate Analysis*, Oxford, UK: Oxford University Press. Cited on p. 95.

Krzanowski, W. J. (2010), "Canonical correlation," in *Encyclopaedic Companion to Medical Statistics*, eds. B. Everitt and C. Palmer, Chichester, UK: John Wiley & Sons, 2nd edition. Cited on p. 96, 99, 101.

Lawley, D. and Maxwell, A. (1963), *Factor Analysis as a Statistical Method*, London, UK: Butterworths. Cited on p. 142.

Leisch, F. (2010), "Neighborhood graphs, stripes and shadow plots for cluster visualization," *Statistics and Computing*, 20, 457–469. Cited on p. 191.

Leisch, F. and Dimitriadou, E. (2019), flexclust: Flexible Cluster Algorithms, URL http://CRAN.R-project.org/package=flexclust, R package version 1.3-1. Cited on p. 194.

Ligges, U. (2010), scatterplot3d: 3D Scatter Plot, URL http://CRAN.R-project.org/package=scatterplot3d, R package version 0.3-31. Cited on p. 47.

Little, R. (2005), "Missing data," in *Encyclopedia of Statistics in Behavioral Science*, eds. B. S. Everitt and D. Howell, Chichester, UK: John Wiley & Sons, pp. 1234–1238. Cited on p. 6.

Little, R. and Rubin, D. (1987), *Statistical Analysis with Missing Data*, New York: John Wiley & Sons. Cited on p. 6.

Longford, N. T. (1993), *Random Coefficient Models*, Oxford, UK: Oxford University Press. Cited on p. 235.

MacCullum, R. C., Browne, M. W., and Sugawara, H. M. (1996), "Power analysis and determination of ssample size for covariance structure models," *Psychological Methods*, 1, 130–149. Cited on p. 222.

MacDonnell, W. R. (1902), "On criminal anthropometry and the identification of criminals," *Biometrika*, 1, 177–227. Cited on p. 102.

Marchini, J. L., Heaton, C., and Ripley, B. D. (2010), fastICA: FastICA Algorithms to perform ICA and Projection Pursuit, URL http://CRAN. R-project.org/package=fastICA, R package version 1.1-13. Cited on p. 102.

Marcoulides, G. A. (2005), "Structural equation modeling: Nontraditional alternatives," in *Encyclopedia of Statistics in Behavioral Science*, eds. B. S. Everitt and D. Howell, Chichester, UK: John Wiley & Sons. Cited on p. 222.

Mardia, K. V., Kent, J. T., and Bibby, J. M. (1979), *Multivariate Analysis*, London, UK: Academic Press. Cited on p. 64, 65, 95, 109, 142, 146, 205.

Marin, J. M., Mengersen, K., and Roberts, C. (2005), "Bayesian modelling and inferences on mixtures of distributions," in *Bayesian Thinking, Modeling and Computation*, eds. D. Dey and C. R. Rao, Amsterdam: Elsevier, pp. 15840–15845. Cited on p. 186, 188.

Marriott, F. (1974), *The Interpretation of Multiple Observations*, London, UK: Academic Press. Cited on p. 88.

McLachlan, G. and Basford, K. (1988), *Mixture Models: Inference and Applications to Clustering*, New York: Marcel Dekker Inc. Cited on p. 186.

McLachlan, G. and Peel, D. (2000), *Finite Mixture Models*, New York: John Wiley & Sons. Cited on p. 186, 188.

Meghani, S. H., Lee, C. S., Hanlon, A. L., and Bruner, D. W. (2009), "Latent class cluster analysis to understand heterogeneity in prostate cancer treatment utilities," *BMC Medical Informatics and Decision Making*, 9. Cited on p. 186.

Meyer, W. and Bendig, A. (1961), "A longitudinal study of the primary mental abilities test," *Journal of Educational Psychology*, 52, 50–60. Cited on p. 223.

Morrison, D. (1990), *Multivariate Statistical Methods*, New York: McGraw-Hill, 3rd edition. Cited on p. 64.

Murray, G. D. and Findlay, J. G. (1988), "Correcting for bias caused by dropouts in hypertension trials," *Statistics in Medicine*, 7, 941–946. Cited on p. 254.

Muthen, B. O. (1978), "Contributions to factor analysis of dichotomous variables," *Psychometrika*, 43, 551–560. Cited on p. 159.

Muthen, B. O. (1984), "A general structural equation model with dichotomous, ordered categorical, and continuous latent variable indicators," *Psychometrika*, 49, 115–132. Cited on p. 222.

Muthen, L. K. and Muthen, B. O. (2002), "How to use a Monte Carlo study to decide on sample size and determine power," *Structural Equation Modeling*, 9, 599–620. Cited on p. 222.

Needham, R. M. (1965), "Computer methods for classification and grouping," in *The Use of Computers in Anthropology*, ed. J. Hymes, The Hague: Mouton, pp. 345–356. Cited on p. 164.

Osborne, J. W. and Costello, A. B. (2004), "Sample size and subject to item ratio in principal components analysis," *Practical Assessment, Research &*

Evaluation, 9, URL `http://www.pareonline.net/getvn.asp?v=9&n=11`. Cited on p. 93.

Paradis, E., Bolker, B., Claude, J., Cuong, H. S., Desper, R., Durand, B., Dutheil, J., Gascuel, O., Heibl, C., Lawson, D., Lefort, V., Legendre, P., Lemon, J., Noel, Y., Nylander, J., Opgen-Rhein, R., Schliep, K., Strimmer, K., and de Vienne, D. (2010), ape: Analyses of Phylogenetics and Evolution, URL `http://CRAN.R-project.org/package=ape`, R package version 2.6-2. Cited on p. 121.

Pearson, K. (1901), "LIII. on lines and planes of closest fit to systems of points in space," *Philosophical Magazine Series 6*, 2, 559–572. Cited on p. 101.

Pett, M., Lackey, N., and Sullivan, J. (2003), *Making Sense of Factor Analysis: The Use of Factor Analysis for Instrument Development in Health Care Research*, Thousand Oaks, CA: Sage Publications. Cited on p. 146.

Pinheiro, J., Bates, D., DebRoy, S., Sarkar, D., and R Development Core Team (2010), nlme: Linear and Nonlinear Mixed Effects Models, URL `http://CRAN.R-project.org/package=nlme`, R package version 3.1-97. Cited on p. 235.

Pinker, S. (1997), *How the Mind Works*, London, UK: The Penguin Press. Cited on p. 163.

Pledger, S. and Phillpot, P. (2008), "Using mixtures to model heterogeneity in ecological capture-recapture studies," *Biometrical Journal*, 50, 1022–1034. Cited on p. 186.

Prim, R. C. (1957), "Shortest connection networks and some generalizations," *Bell System Technical Journal*, 36, 1389–1401. Cited on p. 121.

Proudfoot, J., Goldberg, D., Mann, A., Everitt, B. S., Marks, I., and Gray, J. A. (2003), "Computerized, interactive, multimedia cognitive-behavioural program for anxiety and depression in general practice," *Psychological Medicine*, 33, 217–227. Cited on p. 248.

Rencher, A. (2002), *Methods of Multivariate Analysis*, New York: John Wiley & Sons. Cited on p. 63, 71.

Rencher, A. C. (1995), *Methods of Multivariate Analysis*, New York: John Wiley & Sons. Cited on p. 148.

Rocke, D. and Woodruff, D. (1996), "Identification of outliers in multivariate data," *Journal of the American Statistical Association*, 91, 1047–1061. Cited on p. 23.

Romesburg, H. C. (1984), *Cluster Analysis for Researchers*, Belmont, CA: Lifetime Learning Publications. Cited on p. 123.

Rubin, D. (1976), "Inference and missing data," *Biometrika*, 63, 581–592. Cited on p. 253.

Rubin, D. (1987), *Multiple Imputation for Survey Nonresponse*, New York: John Wiley & Sons. Cited on p. 7.

Rubin, D. B. and Thayer, D. T. (1982), "EM algorithms for ML factor analysis," *Psychometrika*, 47, 69–76. Cited on p. 142.

Sarkar, D. (2008), *Lattice: Multivariate Data Visualization with R*, New York: Springer-Verlag. Cited on p. 51, 52, 57, 180.

Sarkar, D. (2010), lattice: Lattice Graphics, URL http://CRAN.R-project. org/package=lattice, R package version 0.19-13. Cited on p. 51, 180.

Schafer, J. (1999), "Multiple imputation: A primer," *Statistical Methods in Medical Research*, 8, 3–15. Cited on p. 7.

Scheipl, F. (2010), RLRsim: Exact (Restricted) Likelihood Ratio Tests for Mixed and Additive models., URL http://CRAN.R-project.org/ package=RLRsim, R package version 2.0-5. Cited on p. 236.

Scheipl, F., Greven, S., and Küchenhoff, H. (2008), "Size and power of tests for a zero random effect variance or polynomial regression in additive and linear mixed models," *Computational Statistics & Data Analysis*, 52, 3283–3299. Cited on p. 236.

Schmid, C. F. (1954), *Handbook of Graphic Presentation*, New York: Ronald Press. Cited on p. 25.

Schumaker, R. E. and Lomax, R. G. (1996), *A Beginner's Guide to Structural Equation Modeling*, Mahwah, NJ: Lawrence Erlbaum Associates. Cited on p. 202.

Shepard, R. N. (1962a), "The analysis of proximities: Multidimensional scaling with unknown distance function, Part I," *Psychometrika*, 27, 125–140. Cited on p. 121.

Shepard, R. N. (1962b), "The analysis of proximities: Multidimensional scaling with unknown distance function, Part II," *Psychometrika*, 27, 219–246. Cited on p. 121.

Sibson, R. (1979), "Studies in the robustness of multidimensional scaling. Perturbational analysis of classical scaling," *Journal of the Royal Statistical Society, Series B*, 41, 217–229. Cited on p. 109.

Silverman, B. (1986), *Density Estimation*, London, UK: Chapman & Hall/CRC. Cited on p. 42, 43, 46.

Skrondal, A. and Rabe-Hesketh, S. (2004), *Generalized Latent Variable Modeling: Multilevel, Longitudinal and Structural Equation Models*, Boca Raton, FL: Chapman & Hall/CRC. Cited on p. 186, 232, 235, 257.

Sokal, R. R. and Rohlf, F. (1981), *Biometry*, San Francisco, CA: W. H. Freeman, 2nd edition. Cited on p. .

Spearman, C. (1904), "General intelligence objectively determined and measured," *The American Journal of Psychology*, 15, 201–293. Cited on p. 136.

Stanley, W. and Miller, M. (1979), "Measuring technological change in jet fighter aircraft," Technical Report R-2249-AF, Rand Corporation, Santa Monica, CA. Cited on p. 171.

Steinley, D. (2008), "Stability analysis in k-means clustering," *British Journal of Mathematical and Statistical Psychology*, 61, 255–273. Cited on p. 176.

Thomsen, O. O., Wulff, H. R., Martin, A., and Singer, P. A. (1993), "What do gastronenterologists in Europe tell cancer patients?" *The Lancet*, 341, 473–476. Cited on p. 189.

Thorndike, R. M. (2005), "History of factor analysis: A psychological perspective," in *Encyclopedia of Statistics in Behavioral Science*, eds. B. S. Everitt and D. Howell, Chichester, UK: John Wiley & Sons. Cited on p. 136.

Thurstone, L. L. (1931), "Multiple factor analysis," *Psychology Review*, 39, 406–427. Cited on p. 145.

Titterington, D., Smith, A., and Makov, U. (1985), *Statistical Analysis of Finite Mixture Distributions*, John Wiley & Sons. Cited on p. 186.

Tubb, A., Parker, N. J., and Nickless, G. (1980), "The analysis of Romano-British pottery by atomic absorption spectrophotometry," *Archaeometry*, 22, 153–171. Cited on p. .

Tufte, E. R. (1983), *The Visual Display of Quantitative Information*, Cheshire, CT: Graphics Press. Cited on p. 25, 26.

van Hattum, P. and Hoijtink, H. (2009), "Market segmentation using brand strategy research: Bayesian inference with respect to mixtures of log-linear models," *Journal of Classification*, 26, 297–328. Cited on p. 186.

Vanisma, F. and De Greve, J. P. (1972), "Close binary systems before and after mass transfer," *Astrophysics and Space Science*, 87, 377–401. Cited on p. 46.

Velleman, P. and Wilkinson, L. (1993), "Nominal, ordinal, interval, and ratio typologies are misleading," *The American Statistician*, 47, 65–72. Cited on p. 5.

Venables, W. and Ripley, B. D. (2010), MASS: Support Functions and Datasets for Venables and Ripley's MASS, URL http://CRAN.R-project.org/package=MASS, R package version 7.3-9. Cited on p. 123.

Venables, W. N. and Ripley, B. D. (2002), *Modern Applied Statistics with S*, New York: Springer-Verlag, 4th edition. Cited on p. 46, 123.

Verbyla, A., Cullis, B., Kenward, M., and Welham, S. (1999), "The analysis of designed experiments and longitudinal data by using smoothing splines," *Journal of the Royal Statistical Society, Series C*, 48, 269–311. Cited on p. 50.

Wand, M. and Ripley, B. D. (2010), KernSmooth: Functions for Kernel Smoothing, URL http://CRAN.R-project.org/package=KernSmooth, R package version 2.23-4. Cited on p. 46.

Wand, M. P. and Jones, M. C. (1995), *Kernel Smoothing*, London, UK: Chapman & Hall/CRC. Cited on p. 42.

Wheaton, B., Muthen, B., Alwin, D., and Summers, G. (1977), "Assessing reliability and stability in panel models," *Sociological Methodology*, 7, 84–136. Cited on p. 216.

Wolfe, J. (1970), "Pattern clustering by multivariate mixture analysis," *Multivariate Behavioral Research*, 5, 329–350. Cited on p. 187.

Wright, S. (1934), "The method of path coefficients," *The Annals of Mathematical Statistics*, 5, 161–215. Cited on p. 202.

Young, G. and Householder, A. S. (1938), "Discussion of a set of points in terms of their mutual distances," *Psychometrika*, 3, 19–22. Cited on p. 107.

Zerbe, G. O. (1979), "Randomization analysis of the completely randomized design extended to growth and response curves," *Journal of the American Statistical Association*, 74, 215–221. Cited on p. 226.

Index

Printed in Great Britain
by Amazon.co.uk, Ltd.,
Marston Gate.